SOLAR SYSTEM

READING COMPREHENSION

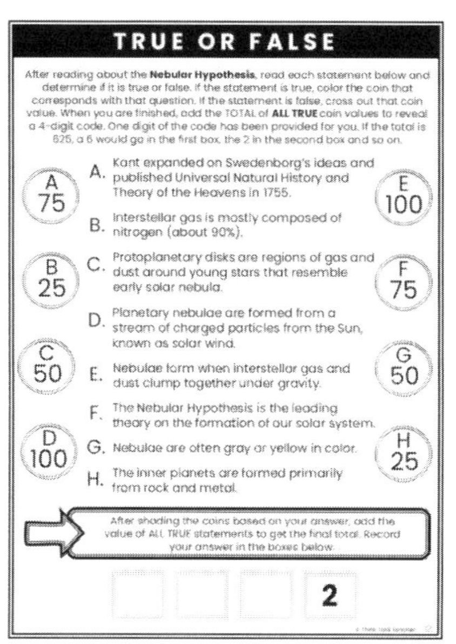

TRUE OR FALSE

After reading about the **Nebular Hypothesis**, read each statement below and determine if it is true or false. If the statement is true, color the coin that corresponds with that question. If the statement is false, cross out that coin value. When you are finished, add the TOTAL of **ALL TRUE** coin values to reveal a 4-digit code. One digit of the code has been provided for you. If the total is 625, a 6 would go in the first box, the 2 in the second box and so on.

A 75 · B 25 · C 50 · D 100 · E 100 · F 75 · G 50 · H 25

A. Kant expanded on Swedenborg's ideas and published Universal Natural History and Theory of the Heavens in 1755.

B. Interstellar gas is mostly composed of nitrogen (about 90%).

C. Protoplanetary disks are regions of gas and dust around young stars that resemble early solar nebula.

D. Planetary nebulae are formed from a stream of charged particles from the Sun, known as solar wind.

E. Nebulae form when interstellar gas and dust clump together under gravity.

F. The Nebular Hypothesis is the leading theory on the formation of our solar system.

G. Nebulae are often gray or yellow in color.

H. The inner planets are formed primarily from rock and metal.

After shading the coins based on your answer, add the value of ALL TRUE statements to get the final total. Record your answer in the boxes below.

			2

MYSTERY WORD

After reading about the **Sun**, determine if each statement below is true or false. Color or shade the boxes of the **TRUE** statements. Next, unscramble the mystery word using the large letters of the **TRUE** statements.

Surrounding the core is the radiative zone. **A**	The average time it takes for sunlight to reach Earth is about forty minutes. **S**	The Sun is predominantly composed of hydrogen and helium. **R**	Solar spicules were first observed by Albert Einstein. **L**
The solar cycle was first suggested in 1838 by astronomer Samuel Schwabe. **P**	The total energy the Sun emits is measured as solar luminosity. **E**	The Sun is so large that it makes up 99.86% of all the mass in our solar system. **T**	Photons produced in the core collide with particles in the dense radiative zone. **R**
The Sun is a green dwarf star (Y-type main-sequence star). **B**	The Sun is composed of five inner layers and two outer layers. **C**	The Sun is a giant ball of hot gas. **E**	Every second, the Sun converts about 4.2 million metric tons of matter into energy. **M**
The Sun experiences a three-year cycle of activity, known as the solar cycle. **D**	The photosphere is the Sun's visible surface, emitting the light we see on Earth. **U**	Solar flares are intense bursts of radiation caused by magnetic energy release. **E**	Sunspots are dark, cooler areas on the Sun's surface caused by magnetic activity. **T**

Unscramble the word using the large bold letters of only the **TRUE** statements.

MULTIPLE CHOICE

After reading about the **Phases of the Moon**, answer each multiple-choice question below. Then, count the number of times you used each letter as an answer (ABCD) to reveal a 4-digit code. Letters may be used more than once or not at all. If a letter option is not used, put a zero in that box.

1. What percent of the Moon is visible during the First Quarter phase?
   A. 10%
   B. 25%
   C. 50%
   D. 65%

2. In terms of moons in our solar system, where does Earth's moon rank?
   A. Second largest moon
   B. Third largest moon
   C. Fourth largest moon
   D. Fifth largest moon

3. Due to gravity, something on the Moon will be how many times lighter than on the Earth?
   A. Six
   B. Twelve
   C. Eighteen
   D. Thirty

4. What is the Moon's outermost layer composed of?
   A. Iron
   B. Calcium
   C. Magnesium
   D. All of the above

5. What is the time it takes the Moon to make a complete cycle of phases known as?
   A. Lunar month
   B. Lunar albedo
   C. Lunar day
   D. None of the above

6. Which phase occurs three or four nights after a New Moon?
   A. Third Quarter
   B. Waning Gibbous
   C. Waxing Crescent
   D. Blue Moon

7. Which ancient culture developed lunar calendars based on the phases of the Moon?
   A. Chinese
   B. Greeks
   C. Babylonians
   D. All of the above

8. How often does a Full Moon occur?
   A. About every 10 days
   B. About every 20 days
   C. About every 30 days
   D. About every 40 days

Count how many times you used each letter as a correct answer (ABCD) to determine the 4-digit code. Record your answer in the boxes below.

# of A's · # of B's · # of C's · # of D's

MYSTERY MATCH

After reading about **Earth**, draw a line from the left-hand column to make a match in the right-hand column. Your line should go through **ONE** letter. When you complete all the matches, use the letters with a line THROUGH them to unscramble a mystery word. You MUST start and end your line at the **arrow points**.

Equinox	B	Earth's size rank compared to other planets
Mantle	C, T	Number of continents on Earth
Blue Planet	S, A	Lowest point on Earth
Mariana Trench	I	Earth's middle layer
Fifth Largest	T, O	Contains the ozone layer
Seven	H, R	Equal day and night
Stratosphere	L, D	Ground or soil
Erda	N, O	Earth's nickname

Unscramble the 8 letters to reveal a mystery word:

DOUBLE PUZZLE

After reading about **Mars**, determine the word that corresponds with the statements provided below. Spell the corresponding word in the boxes to the right. You may or may not use all squares provided for each answer. Any numerical answers must be spelled out. Next, use the numbers **under** indicated letters to reveal a secret word.

1. The longest season on Mars (7)
2. The moon of Mars that is closest to the planet (5)
3. ___ Manneris is the biggest crater in the solar system (4)
4. Mars has an ___ tilt of about 25 degrees (2)
5. Phobos goes around Mars ___ times in one Martian day (1)
6. Babylonians referred to the planet as this name for the god of war (8)
7. Reddish Martian soil
8. NASA'S first successful touch down on the surface of Mars was with the ___ lander (9)
9. The highest mountain in the solar system is ___ Mons (3)
10. Number of minutes it takes sunlight to reach Mars (6)

SECRET WORD: 1 2 3 4 5 6 7 8 9

PARAGRAPH CODE

After reading about the **Lunar Eclipse**, head back to the reading and number ALL the paragraphs in the reading passage. Then, read each statement below and determine which paragraph **NUMBER** the statement can be found in. Paragraph numbers MAY be used more than one time or not at all. Follow the directions below to reveal the 4-digit code.

A. Ending phases of eclipses occur when the sequence reverses as the Moon gradually exits the umbra and then the penumbra.

B. During the Middle Ages, lunar eclipses were often seen as predictions of natural disasters.

C. A Maximum Eclipse is the midpoint of the total eclipse, when the Moon is at its deepest within Earth's shadow.

D. Penumbral lunar eclipses are relatively common, occurring about one to four times per year.

E. Depending on the alignment, a lunar eclipse can be partial, total, or penumbral.

F. Lunar eclipses are visible from anywhere on Earth where the Moon is above the horizon during the eclipse.

G. Historically, these celestial phenomena have been tied to mythology, folklore, and religion, often interpreted as omens of significant events.

H. A partial eclipse is when part of the Moon enters the umbra.

ELIMINATE ALL EVEN-NUMBERED paragraphs that you used as an answer. Record the remaining numbers (in the SAME order in which you recorded them above) in the boxes below.

SECRET CODES & MYSTERY WORDS

By Lisa Fink

Contact the author :
Lisa@ThinkTankTeacher.com

ABOUT THIS WORKBOOK

Science doesn't have to be boring — especially when it comes with a dash of mystery and a twist of F-U-N, including secret codes and mystery words! This collection of **21** reading comprehension passages is designed to spark curiosity and make homeschool **science** curriculum come alive for kids ages 10+.

WHAT'S INSIDE?

21 Reading Passages: (2 full pages each)
Each passage is crafted to draw your kids in, making science relatable and intriguing for learners at home or in the classroom. To ensure understanding, each reading comprehension passage comes with two interactive worksheets.

Worksheet 1: Crack the 4-Digit Code!
This activity turns reading into a puzzle, keeping kids motivated to dive deeper into the material. Your learners will answer comprehension questions based on the passage and correct answers will reveal a secret 4-digit code. Programmable locks aren't required, but they can take the fun to the next level!

Worksheet 2: Mystery Word!
To ensure they've truly grasped the material, kids answer another set of questions, based on the same reading passage, that unveil a secret word.

WHY IT WORKS:

Active Engagement: By turning reading comprehension into a game, kids are more likely to stay engaged, motivated to learn and eager to finish the task. Game-based learning requires active participation. Rather than passively reading and answering questions, kids are actively seeking answers to solve the puzzle.

Encourages Critical Thinking: Game elements often require kids to think critically and make connections between different pieces of information. To find the correct answers and solve the puzzle, they must analyze the text carefully, improving their comprehension skills.

Reinforcement Through Fun: The excitement of uncovering codes and secret words reinforces key scientific concepts in a way that sticks. Kids may be more willing to go back and re-read sections of the passage to ensure they have the correct answers.

Incentive to Learn: The reward of discovering a secret code or mystery word acts as a powerful incentive. Kids are more likely to pay close attention to details in the reading passage, knowing that each answer brings them closer to solving the puzzle. This can lead to a deeper understanding of the content.

Increased Motivation: When an assignment includes elements like secret codes or mystery words, it feels less like work and more like a fun challenge. This added layer of excitement motivates kids to complete the task to "win" or uncover the hidden element.

Versatile Use: These activities are perfect for homeschooling or classrooms — use as part of your daily lessons, supplemental homework, or independent study.

Promotes Positive Reinforcement: Successfully uncovering a secret code or mystery word provides immediate positive feedback, which reinforces the learning experience. This sense of achievement boosts confidence and encourages kids to tackle future activities with enthusiasm.

HOW TO FIND THE MYSTERY WORD

There are three different types of mystery word activities.

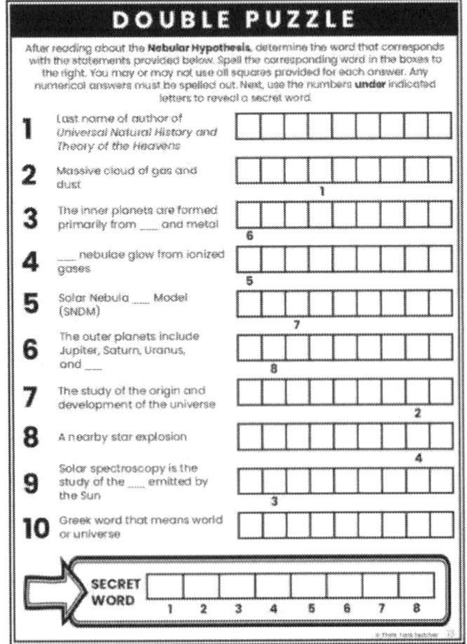

DOUBLE PUZZLE

After reading about the **Nebular Hypothesis**, determine the word that corresponds with the statements provided below. Spell the corresponding word in the boxes to the right. You may or may not use all squares provided for each answer. Any numerical answers must be spelled out. Next, use the numbers **under** indicated letters to reveal a secret word.

1 Last name of author of *Universal Natural History and Theory of the Heavens*

2 Massive cloud of gas and dust

3 The inner planets are formed primarily from ___ and metal

4 ___ nebulae glow from ionized gases

5 Solar Nebula ___ Model (SNDM)

6 The outer planets include Jupiter, Saturn, Uranus, and ___

7 The study of the origin and development of the universe

8 A nearby star explosion

9 Solar spectroscopy is the study of the ___ emitted by the Sun

10 Greek word that means world or universe

SECRET WORD 1 2 3 4 5 6 7 8

DOUBLE PUZZLE

After reading the passage, determine the correct word that corresponds with statements 1-10. Spell the corresponding word in the boxes to the right. You may or may not use all squares provided for each answer. Any numerical answers **must** be spelled out. Next, use the numbers **under** indicated letters to reveal a secret word.

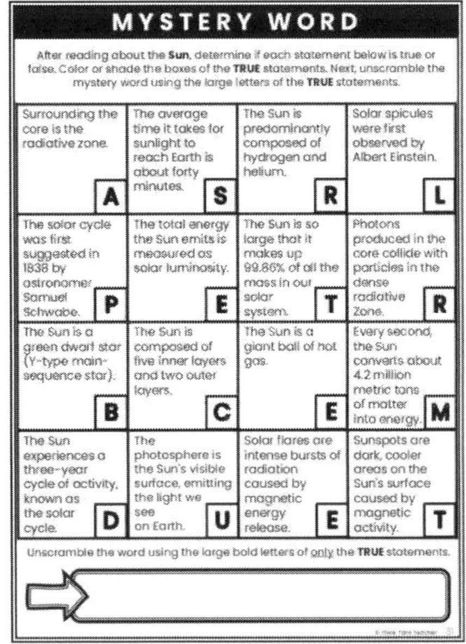

MYSTERY WORD

After reading about the **Sun**, determine if each statement below is true or false. Color or shade the boxes of the **TRUE** statements. Next, unscramble the mystery word using the large letters of the **TRUE** statements.

Surrounding the core is the radiative zone. **A**	The average time it takes for sunlight to reach Earth is about forty minutes. **S**	The Sun is predominantly composed of hydrogen and helium. **R**	Solar spicules were first observed by Albert Einstein. **L**
The solar cycle was first suggested in 1838 by astronomer Samuel Schwabe. **P**	The total energy the Sun emits is measured as solar luminosity. **E**	The Sun is so large that it makes up 99.86% of all the mass in our solar system. **T**	Photons produced in the core collide with particles in the dense radiative zone. **R**
The Sun is a green dwarf star (Y-type main-sequence star). **B**	The Sun is composed of five inner layers and two outer layers. **C**	The Sun is a giant ball of hot gas. **E**	Every second, the Sun converts about 4.2 million metric tons of matter into energy. **M**
The Sun experiences a three-year cycle of activity, known as the solar cycle. **D**	The photosphere is the Sun's visible surface, emitting the light we see on Earth. **U**	Solar flares are intense bursts of radiation caused by magnetic energy release. **E**	Sunspots are dark, cooler areas on the Sun's surface caused by magnetic activity. **T**

Unscramble the word using the large bold letters of only the **TRUE** statements.

MYSTERY WORD

After reading the passage, determine if each statement is true or false. Color or shade the boxes of the **TRUE** statements. After all TRUE answers have been shaded, unscramble the mystery word using the large, bold letters of only the TRUE statements.

MYSTERY MATCH

After reading about the **Phases of the Moon**, draw a line from the left-hand column to make a match in the right-hand column. Your line should go through **ONE** letter. When you complete all the matches, use the letters with a line THROUGH them to unscramble a mystery word. You MUST start and end your line at the **arrow points**.

Waxing · Dark Moon · Satellite · Blue Moon · Ellipse · First Quarter · Waning · Selene

J D E F E B K N L E T O P C G H R

Second Full Moon in a month · Another name for Half Moon · When the Moon appears to be getting bigger · Moon goddess of Greek mythology · Another name for New Moon · When the Moon appears to be getting smaller · A celestial body that orbits a planet · Not a perfect circle

Unscramble the 8 letters to reveal a mystery word:

MYSTERY MATCH

After reading the passage, draw a line from the left-hand column to make a match in the right-hand column. Your line should go THROUGH **one** letter. When you complete all the matches, use the letters with a line THROUGH them to unscramble a mystery word. You MUST start and end your line at the arrow points. Then, unscramble the 8 letters to reveal the mystery word. **IMPORTANT: Your line MUST start with an arrow TIP on the left and end at an arrow TIP on the right.** If you do not draw from arrow tip to arrow tip, you will not be able to reveal the hidden word.

HOW TO FIND THE SECRET CODE

There are three different types of secret code activities.

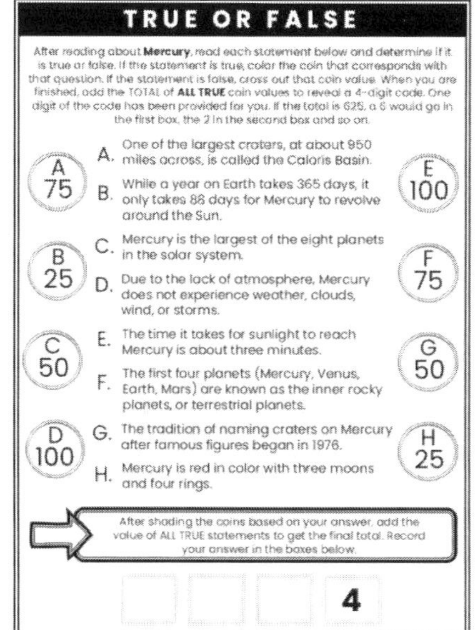

TRUE OR FALSE

After reading about **Mercury**, read each statement below and determine if it is true or false. If the statement is true, color the coin that corresponds with that question. If the statement is false, cross out that coin value. When you are finished, add the TOTAL of **ALL TRUE** coin values to reveal a 4-digit code. One digit of the code has been provided for you. If the total is 625, a 6 would go in the first box, the 2 in the second box and so on.

A 75 — A. One of the largest craters, at about 950 miles across, is called the Caloris Basin.

B 25 — B. While a year on Earth takes 365 days, it only takes 88 days for Mercury to revolve around the Sun.

C 50 — C. Mercury is the largest of the eight planets in the solar system.

D 100 — D. Due to the lack of atmosphere, Mercury does not experience weather, clouds, wind, or storms.

E 100 — E. The time it takes for sunlight to reach Mercury is about three minutes.

F 75 — F. The first four planets (Mercury, Venus, Earth, Mars) are known as the inner rocky planets, or terrestrial planets.

G 50 — G. The tradition of naming craters on Mercury after famous figures began in 1976.

H 25 — H. Mercury is red in color with three moons and four rings.

After shading the coins based on your answer, add the value of ALL TRUE statements to get the final total. Record your answer in the boxes below.

			4

TRUE OR FALSE

After reading the passage, determine if each statement is true or false. If the statement is true, color, circle, or shade-in the coin that corresponds with that question. If the statement is false, cross out that coin value. When you are finished, add the **TOTAL** of **ALL TRUE** coin values to reveal a 4-digit code. If you are working on statement D, be sure you find the coin that is labeled D. The final number of the code will always be provided for you.

PARAGRAPH CODE

After reading about **Jupiter**, head back to the reading and number ALL the paragraphs in the reading passage. Then, read each statement below and determine which paragraph **NUMBER** the statement can be found in. Paragraph numbers MAY be used more than one time or not at all. Follow the directions below to reveal the 4-digit code.

A Jupiter has an interesting "Great Red Spot" that Giovanni Cassini, an Italian astronomer, spotted in 1665.

B Jupiter was named after the king of the Roman gods in ancient mythology.

C Jupiter has four faint rings called the Halo ring, Main ring, Amalthea Gossamer ring, and the Thebe Gossamer ring.

D This discovery provided evidence against the geocentric model (Earth-centered universe).

E Ganymede is the ninth largest object and the largest moon in the solar system (larger than Mercury).

F You could fit eleven Earths side by side within Jupiter's diameter.

G Jupiter is a gas planet, like Saturn, Uranus, and Neptune, made up of 90% hydrogen and 10% helium.

H Winds within the Great Red Spot can reach speeds of up to four hundred miles per hour.

ELIMINATE ALL EVEN-NUMBERED paragraphs that you used as an answer. Record the remaining numbers (in the SAME order in which you recorded them above) in the boxes below.

PARAGRAPH CODE

After reading the passage, number **ALL** the paragraphs. Then, read each statement and determine which paragraph **NUMBER** the statement can be found in. Paragraph numbers MAY be used more than one time or not at all. When all paragraph numbers have been found, **ELIMINATE ALL EVEN-NUMBERED** paragraphs that you used as an answer. Record the remaining numbers (in the SAME order in which you recorded them) in the boxes at the bottom. **Example: 85362147 = 5317 (after crossing off all EVEN numbers, the remaining numbers in the order in which they were found is 5317)**

MULTIPLE CHOICE

After reading about **NASA**, answer each multiple-choice question below. Then, count the number of times you used each letter as an answer (ABCD) to reveal a 4-digit code. Letters may be used more than once or not at all. If a letter option is not used, put a zero in that box.

1 How many total flights with astronauts did the Mercury program consist of?
A. Six
B. Ten
C. Twelve
D. Fifteen

2 In 1969, what was the first mission in the world to successfully put a man on the Moon?
A. Friendship 7
B. Agena 3
C. Vostok I
D. Apollo 11

3 Which president started the National Advisory Committee for Aeronautics (NACA) in 1915?
A. John F. Kennedy
B. Donald Trump
C. Dwight D. Eisenhower
D. Woodrow Wilson

4 How many times did Glenn orbit the Earth aboard the Friendship 7?
A. Three
B. Five
C. Seven
D. Nine

5 When was NASA established by President Dwight D. Eisenhower?
A. 1958
B. 1963
C. 1968
D. 1971

6 The first successful manned mission of NASA was part of which project?
A. Project Agena
B. Project Apollo
C. Project Gemini
D. Project Mercury

7 In 1961, which president announced that America should put a man on the Moon?
A. John F. Kennedy
B. Donald Trump
C. Dwight D. Eisenhower
D. Woodrow Wilson

8 Russian astronaut Yuri Gagarin become the first man to orbit the Earth aboard what spacecraft?
A. Apollo 11
B. Freedom 7
C. Vostok I
D. None of the above

Count how many times you used each letter as a correct answer (ABCD) to determine the 4-digit code. Record your answer in the boxes below.

# of A's	# of B's	# of C's	# of D's

MULTIPLE CHOICE

After reading the passage, answer each multiple-choice question. Then, count how many times you used each letter as a correct answer (ABCD) to determine the 4-digit code. Letters may be used more than once or not at all. If a letter option is not used, put a zero in that box. **Example: B,C,C,D,C,A,A,C = 2141** (The A was chosen as a correct answer for 2 questions, the B was chosen 1 time, C was used 4 times and D was chosen as a correct answer 1 time)

ORIGINS AND EARLY ASTRONOMERS

NEBULAR HYPOTHESIS

Cosmogony is the study of the origin and development of the universe, especially the solar system. The term comes from the Greek words kosmos (world or universe) and gonos (creation). It addresses questions about how the universe began, the processes that shaped it, and its ultimate structure.

The Nebular Hypothesis is the leading theory on the formation of our solar system. The theory was originally proposed by German philosopher Immanuel Kant and later refined by French mathematician Pierre-Simon Laplace. This possible explanation suggests that our solar system formed from a massive cloud of gas and dust, called a nebula. Over millions of years, gravity caused the nebula to collapse and spin, flattening into a disk with the Sun forming at its center.

Origins of the Nebular Hypothesis

Around 1734, Emanuel Swedenborg, a Swedish scientist, contributed early ideas to what would later develop into the "Nebular Hypothesis" for the origin of the solar system.

Immanuel Kant proposed that the Sun and planets originated from a rotating cloud of gas. Kant expanded on Swedenborg's ideas and published *Universal Natural History and Theory of the Heavens* in 1755. Laplace later expanded on Kant's model, publishing *Exposition du Systeme du Monde* in 1796. Laplace suggested that as the cloud spun, it flattened and condensed due to gravity, forming rings that eventually merged into planets. This early hypothesis laid a foundation, but the lack of scientific tools at the time left gaps in verifying or proving these ideas.

Likely Formation Process

About 4.6 billion years ago, the Sun and planets formed from a large cloud of gas and dust. Likely, a shock wave from a nearby star explosion, or supernova, caused the cloud to collapse. The Sun took shape at the center, while planets formed in a flat, spinning disk around it. Moons also formed in a similar way around giant planets. As the Sun ignited, its strong solar wind cleared away gas and dust, leaving rocky asteroids and icy comets as leftovers.

As gravity drew heavier elements toward the center of the solar nebula, the inner planets (Mercury, Venus, Earth, and Mars) formed primarily from rock and metal. These materials could withstand the higher temperatures close to the Sun. In contrast, the outer planets (Jupiter, Saturn, Uranus, and

Neptune) formed farther away in cooler regions of the solar system. They condensed from lighter materials like hydrogen, helium, water, ammonia, and methane. In these colder areas, such materials could freeze into solid particles, forming the cores of gas giants.

Advances in Understanding

In the mid-1900s, astronomers such as Gerard Kuiper and Harold Urey advanced the idea that the solar system's current structure could be explained by a rotating, collapsing disk. This idea received strong support from observations in protoplanetary disks - regions of gas and dust around young stars that resemble early solar nebula. Observations from telescopes like Hubble and ALMA have captured images of disks around young stars, reinforcing the nebular model.

Nebulae form when interstellar gas and dust clump together under gravity. Interstellar gas is the collection of gases found between stars in a galaxy. This gas is mostly hydrogen (about 90%) and helium (about 9%), with small amounts of heavier elements like carbon, oxygen, and nitrogen. As these materials gather, they can eventually collapse under their own weight, leading to the formation of new stars and solar systems.

Nebulae come in different types, such as emission nebulae (glowing from ionized gases), reflection nebulae (reflecting light from nearby stars), and planetary nebulae (formed from dying stars). In emission nebulae, ionized hydrogen gas glows brightly because it is energized by nearby stars. This gives nebulae their distinctive colors, often red or green, depending on the gases involved. The Orion Nebula, which is part of the constellation Orion, is visible from Earth.

Solar spectroscopy is the study of the light emitted by the Sun to understand its composition, structure, and physical processes. In the 1970s, research into solar spectroscopy offered more evidence that the Sun and planets share a similar composition. The research also supported the theory that they came from the same material.

Modern Interpretations and Adjustments

The Nebular Hypothesis has evolved with new data and is now part of the Solar Nebula Disk Model (SNDM). This model explains the diversity of planetary compositions and positions. The model also suggests that solar winds, gravitational interactions, and migration of early protoplanets influenced the current layout of the solar system. The SNDM was largely based on research from Soviet astronomer Victor Safronov.

TRUE OR FALSE

After reading about the **Nebular Hypothesis**, read each statement below and determine if it is true or false. If the statement is true, color the coin that corresponds with that question. If the statement is false, cross out that coin value. When you are finished, add the TOTAL of **ALL TRUE** coin values to reveal a 4-digit code. One digit of the code has been provided for you. If the total is 625, a 6 would go in the first box, the 2 in the second box and so on.

A. Kant expanded on Swedenborg's ideas and published Universal Natural History and Theory of the Heavens in 1755.

B. Interstellar gas is mostly composed of nitrogen (about 90%).

C. Protoplanetary disks are regions of gas and dust around young stars that resemble early solar nebula.

D. Planetary nebulae are formed from a stream of charged particles from the Sun, known as solar wind.

E. Nebulae form when interstellar gas and dust clump together under gravity.

F. The Nebular Hypothesis is the leading theory on the formation of our solar system.

G. Nebulae are often gray or yellow in color.

H. The inner planets are formed primarily from rock and metal.

After shading the coins based on your answer, add the value of ALL TRUE statements to get the final total. Record your answer in the boxes below.

DOUBLE PUZZLE

After reading about the **Nebular Hypothesis**, determine the word that corresponds with the statements provided below. Spell the corresponding word in the boxes to the right. You may or may not use all squares provided for each answer. Any numerical answers must be spelled out. Next, use the numbers **under** indicated letters to reveal a secret word.

#	Statement	Boxes (9)	Indicated letter
1	Last name of author of *Universal Natural History and Theory of the Heavens*		
2	Massive cloud of gas and dust		1 (under box 4)
3	The inner planets are formed primarily from ___ and metal		6 (under box 1)
4	___ nebulae glow from ionized gases		5 (under box 1)
5	Solar Nebula ___ Model (SNDM)		7 (under box 3)
6	The outer planets include Jupiter, Saturn, Uranus, and ___		8 (under box 2)
7	The study of the origin and development of the universe		2 (under box 8)
8	A nearby star explosion		4 (under box 8)
9	Solar spectroscopy is the study of the ___ emitted by the Sun		3 (under box 2)
10	Greek word that means world or universe		

SECRET WORD

1	2	3	4	5	6	7	8

NICOLAUS COPERNICUS

Nicolaus Copernicus was a Polish astronomer who reshaped scientific thought and paved the way for future astronomers to further develop our understanding of the solar system. His work changed how people understood the universe by introducing the idea that the Sun, not Earth, was at the center of our solar system. This idea was revolutionary because, for centuries, people believed Earth was at the center of everything.

Early Life and Education

Copernicus was born on February 19, 1473, in Toruń, Poland. He was the youngest of four children born to a merchant family. When Copernicus was just ten years old, his father passed away. His uncle, Lucas Watzenrode, took charge of his upbringing and made sure that Copernicus received a top-notch education.

In 1491, Copernicus enrolled at the University of Kraków where he studied mathematics, astronomy, and philosophy. He later continued his education at the University of Bologna in Italy, where he studied canon (religious) law. It was during his time in Bologna that Copernicus began developing an interest in challenging the current system of astronomy.

Geocentric Model

For almost 2,000 years, people in Europe believed in ideas about the universe that came from ancient Greek scientists like Aristotle and Ptolemy. They thought that Earth was the center of everything, and that the Sun, Moon, and planets all moved around Earth. This was called the geocentric model, and it was considered a fact for many centuries.

The geocentric model was accepted by most people in Europe, including the Catholic Church. It seemed to make sense at the time, because it matched what people could see in the sky. They thought that Earth was the most important place in the universe. This idea fit well with everyday observations and seemed to match the teachings of the Bible, which many people relied on for understanding the world.

The Heliocentric Model

Copernicus proposed an alternative theory: the heliocentric model. The word "helios" is an ancient Greek word meaning "sun." This theory suggested that the Sun was stationary at the center of the universe, with Earth and the other planets revolving around it. Copernicus argued that Earth rotated on its axis daily and orbited the Sun annually.

At the time Copernicus proposed his idea, there were no telescopes to directly observe planets' movements in detail. The evidence Copernicus provided was mathematical and theoretical, but not physical. This made it difficult for people to see the truth of his idea.

Development of the Theory

Between 1501 and 1503, Copernicus studied at the University of Padua. He returned to Poland in 1503. Despite his church duties in Poland, Copernicus continued to refine his astronomical observations.

Around 1514, he wrote a manuscript called *Commentariolus* ("Little Commentary"), which outlined the basic principles of his heliocentric theory. His book *De Revolutionibus Orbium Coelestium* (On the Revolutions of the Celestial Spheres) consisted of six volumes. It was completed around 1532 but was not published until 1543, shortly before his death. Copernicus hesitated to publish the work, likely fearing backlash from the Church and academic circles. However, German mathematician Georg Rheticus persuaded him to release it.

Impact and Legacy

The heliocentric theory initially faced resistance. Many scholars dismissed it as contrary to common sense and religious doctrine. The Catholic Church eventually placed *De Revolutionibus* on its Index of Forbidden Books in 1616. The Church felt that the idea went against certain Bible verses that seemed to say the Earth was the center of the universe. The ban on the book was not lifted until 1835.

Despite this, the work of Copernicus greatly influenced later scientists. Tycho Brahe (1546–1601) expanded upon his observations. Johannes Kepler (1571–1630) refined the heliocentric model by demonstrating that planetary orbits were elliptical rather than circular. Galileo Galilei (1564–1642) provided evidence using telescopic observations. Isaac Newton (1643–1727) later unified these findings with his laws of motion and gravitation.

Copernicus died on May 24, 1543, at the age of seventy. The insights of Copernicus sparked the Scientific Revolution, influencing generations of scientists and thinkers. It was not until Isaac Newton came up with the Law of Universal Gravitation that the heliocentric view became widely accepted. This law explained how gravity works to pull objects toward each other. It showed that the planets stay in orbit around the Sun because of the Sun's much stronger gravitational pull. Newton's work helped support the idea that the Sun, not Earth, was at the center of the solar system.

PARAGRAPH CODE

After reading about **Nicolaus Copernicus**, head back to the reading and number ALL the paragraphs in the reading passage. Then, read each statement below and determine which paragraph **NUMBER** the statement can be found in. Paragraph numbers MAY be used more than one time or not at all. Follow the directions below to reveal the 4-digit code.

A The geocentric model was accepted by most people in Europe, including the Catholic Church.

B It was completed around 1532 but was not published until 1543, shortly before his death.

C Between 1501 and 1503, Copernicus studied at the University of Padua.

D It was not until Isaac Newton came up with the Law of Universal Gravitation that the heliocentric view became widely accepted.

E The evidence Copernicus provided was mathematical and theoretical, but not physical.

F Copernicus was born on February 19, 1473, in Toruń, Poland.

G The Catholic Church eventually placed *De Revolutionibus* on its Index of Forbidden Books in 1616.

H In 1491, Copernicus enrolled at the University of Kraków where he studied mathematics, astronomy, and philosophy.

ELIMINATE ALL EVEN-NUMBERED paragraphs that you <u>used</u> as an answer. Record the remaining numbers (in the SAME order in which you recorded them above) in the boxes below.

MYSTERY WORD

After reading about **Nicolaus Copernicus**, determine if each statement below is true or false. Color or shade the boxes of the **TRUE** statements. Next, unscramble the mystery word using the large letters of the **TRUE** statements.

The insights of Copernicus sparked the Scientific Revolution. **R**	Kepler proved that planetary orbits were elliptical rather than circular. **L**	Newton's work supported the idea that the Sun was at the center of the solar system. **A**	Nicolaus Copernicus was a French astronomer. **C**
The word "helios" is an ancient Greek word meaning "moon." **H**	In 1491, Copernicus enrolled at the University of Kraków. **Y**	Copernicus hesitated to publish his work. **E**	*De Revolutionibus Orbium Coelestium* consisted of twelve volumes. **S**
Copernicus died on May 24, 1543, at the age of seventy. **A**	Copernicus could easily provide physical evidence of his theory. **O**	Copernicus argued that Earth rotated on its axis daily and orbited the Sun annually. **P**	Copernicus was born on February 19, 1476, in Madrid, Spain. **B**
Nicolaus was the oldest of four children. **D**	The geocentric model was considered a fact for many centuries. **T**	The ban of *De Revolutionibus* was lifted just three years later. **I**	The heliocentric theory initially faced resistance. **N**

Unscramble the word using the large bold letters of <u>only</u> the **TRUE** statements.

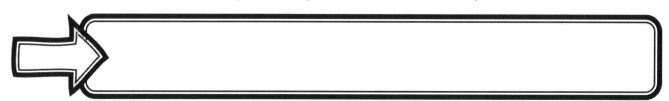

JOHANNES KEPLER

Johannes Kepler (1571–1630) was a German mathematician, astronomer, and key figure in the Scientific Revolution. His groundbreaking discoveries in planetary motion transformed our understanding of the cosmos. His discoveries also laid the groundwork for Isaac Newton's theory of gravitation. Kepler's laws of planetary motion explained the orbits of planets and their speeds, marking a significant advancement over earlier models of the solar system.

Born on December 27, 1571, in Germany, Kepler showed an early interest for mathematics and astronomy. Despite his family's financial struggles, Kepler secured a scholarship to study at the University of Tübingen. His original plan was to study theology (religion) and philosophy. At Tübingen, he was introduced to the heliocentric model proposed by Nicolaus Copernicus. Unlike the geocentric model endorsed by the Catholic Church, which placed Earth at the center of the universe, Copernicus' model theorized that the Sun was at the center, with planets orbiting around it.

Kepler's Work with Tycho Brahe

In 1600, Kepler joined the observatory of Danish astronomer Tycho Brahe in Prague. Brahe spent many years carefully observing the night sky and recording detailed information about how planets, especially Mars, moved through space. After Brahe's death in 1601, Kepler inherited his data and was appointed Imperial Mathematician to Holy Roman Emperor Rudolf II.

Kepler's Laws of Planetary Motion

After years of detailed analysis, Kepler formulated three laws of planetary motion between 1609 and 1619. The first law was known as the Law of Ellipses. Kepler discovered that planets do not move in perfect circles, as previously believed, but rather in elliptical orbits with the Sun at one focus. This was a major shift from the circular orbits assumed by Ptolemaic and Copernican models. The publication of this law in *Astronomia Nova* (1609) was revolutionary and explained the irregular speeds of planets.

The second law, known as the Law of Equal Areas, states that a line segment joining a planet to the Sun sweeps out equal areas in equal time intervals. In practical terms, this means that planets move faster when they are closer to the Sun and slower when they are farther away. This discovery showed that planets do not move at the same speed throughout their orbits, as earlier models had suggested. For example, Earth travels faster in its orbit during perihelion (its closest point to the Sun in January) and slower during aphelion (its farthest point in July).

Kepler's third law, known as the Law of Harmonies, established a relationship between the time a planet takes to orbit the Sun (orbital period) and the average distance from the Sun (semi-major axis). Published in his work *Harmonices Mundi* ("The Harmony of the World") in 1619, this law revealed a universal mathematical order in the cosmos.

Kepler's Influence

Kepler's laws not only supported the heliocentric model but also provided a foundation for Isaac Newton's formulation of the Law of Universal Gravitation in 1687. Newton showed that Kepler's laws were consistent with his own theory, explaining that gravity is the force causing planets to orbit the Sun.

Kepler and the Mechanics of Human Sight

In 1604, Kepler published a treatise on optics titled *Astronomiae Pars* Optica ("The Optical Part of Astronomy"). This groundbreaking work explained how human sight works, including the role of the lens in the eye. Kepler was the first to describe the process by which an image is inverted when it is projected onto the retina. He also detailed how light rays converge through the lens to form an image, paving the way for a scientific understanding of vision.

Improvements to the Telescope

Kepler built upon the earlier invention of the telescope by Dutch opticians and Galileo's improvements. In 1611, Kepler published *Dioptrice*, in which he analyzed how lenses bend light and introduced the idea of the "Keplerian telescope." Unlike Galileo's refracting telescope, which used a concave lens as the eyepiece, Kepler proposed using a convex lens. This adjustment allowed for a wider field of view and greater magnification, although it did invert the image.

Kepler passed away on November 15, 1630, in Germany. Following his death, he was honored for his contributions to understanding planetary motion. A lunar crater was named in his honor, located in the Moon's Mare Insularum region. Discovered in 1929, Asteroid 1134 Kepler, bears his name as a tribute to his revolutionary astronomical work. Launched by NASA in 2009, the Kepler Space Telescope was designed to search for exoplanets. It remained in operation until 2018.

MYSTERY MATCH

After reading about **Johannes Kepler**, draw a line from the left-hand column to make a match in the right-hand column. Your line should go through **ONE** letter. When you complete all the matches, use the letters with a line THROUGH them to unscramble a mystery word. You MUST start and end your line at the **arrow points**.

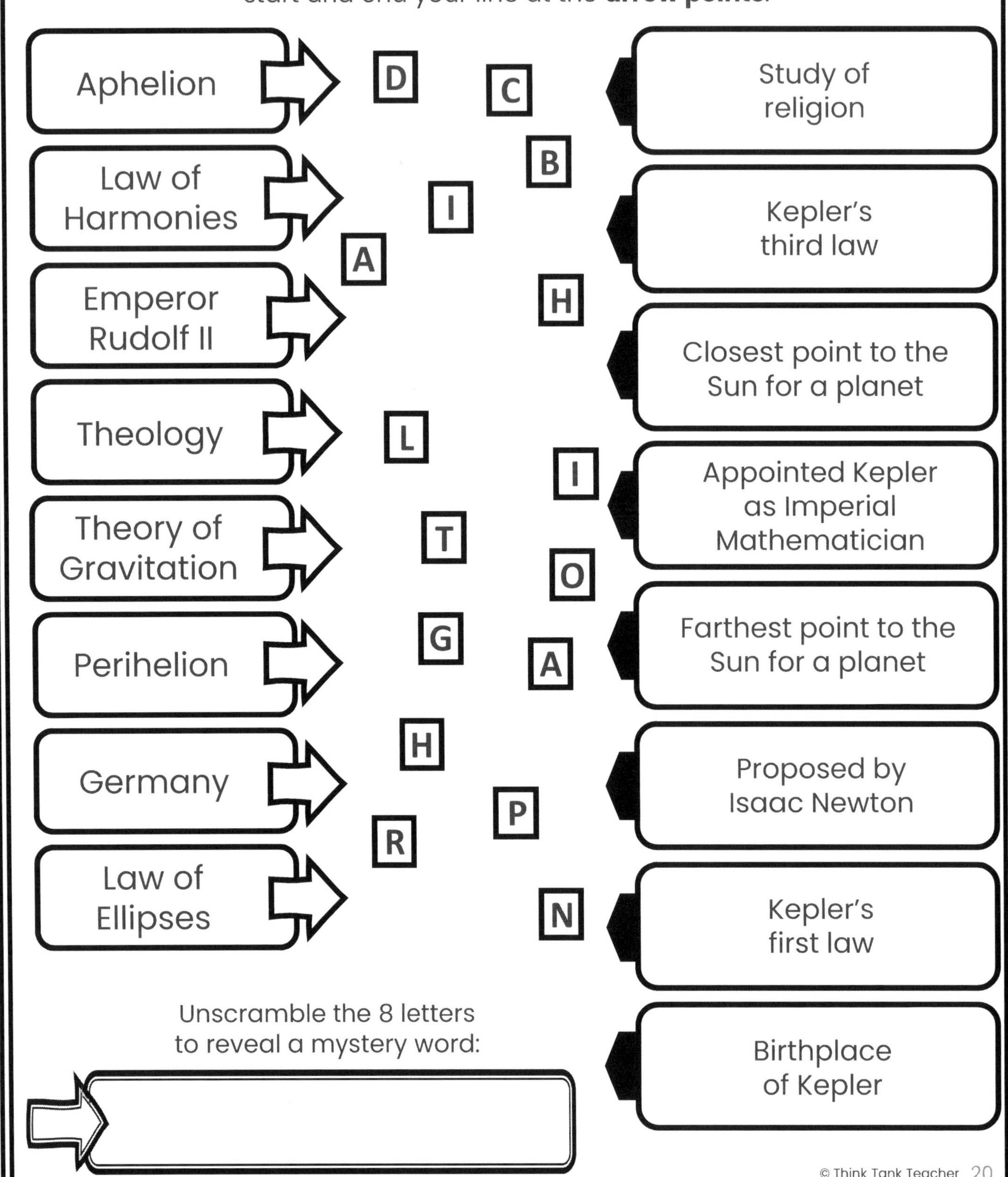

Unscramble the 8 letters to reveal a mystery word:

MULTIPLE CHOICE

After reading about **Johannes Kepler**, answer each multiple-choice question below. Then, count the number of times you used each letter as an answer (ABCD) to reveal a 4-digit code. Letters may be used more than once or not at all. If a letter option is not used, put a zero in that box.

1 How many laws of planetary motion did Kepler formulate between 1609 and 1619?

A. One
B. Two
C. Three
D. Four

2 In which of Kepler's publications did he analyze how lenses bend light?

A. Dioptrice
B. Harmonices Mundi
C. Astronomia Nova
D. None of the above

3 Who proposed the heliocentric model?

A. Isaac Newton
B. Ptolemy
C. Tycho Brahe
D. Nicolaus Copernicus

4 What is a planet's average distance from the Sun known as?

A. Axial tilt
B. Semi-major axis
C. Concave-minor axis
D. None of the above

5 Where did Kepler join the observatory of Danish astronomer Tycho Brahe?

A. Paris
B. Prague
C. London
D. Berlin

6 Where did Kepler secure a scholarship to study?

A. Harvard University
B. University of Tübingen
C. Imperial College
D. University of Cambridge

7 When was *Astronomia Nova* published?

A. 1607
B. 1609
C. 1611
D. 1613

8 Following his death, how was Kepler honored for his contributions?

A. Telescope named after him
B. Lunar crater named after him
C. Asteroid named after him
D. All of the above

Count how many times you used each letter as a correct answer (ABCD) to determine the 4-digit code. Record your answer in the boxes below.

# of A's	# of B's	# of C's	# of D's

GALILEO GALILEI

Galileo Galilei was an astronomer, scientist, mathematician, inventor, and accomplished musician. He was born in Pisa, Italy on February 15, 1564, and grew up during the Italian Renaissance. The Renaissance was a time of rebirth for education, science, art, literature, music, and ultimately, a better life.

Galileo made multiple discoveries through a telescope and published his findings. He also studied and tested the science of physics and motion. He challenged the ideas of ancient philosophers through experimentation.

He was the eldest of six children in a family with noble but modest means. His father, Vincenzo Galilei, was a musician and a scholar who encouraged his son's education.

When Galilei attended the University of Pisa, he studied medicine, hoping to become a doctor. There, he realized he did not want to study medicine anymore and instead found an interest in math and science. He was later forced to leave school due to financial difficulties and got a job as a teacher. As a math teacher, he began to test scientific theories. His observations went against scientific laws of that time.

Discoveries in Physics: The Law of the Pendulum

One of Galileo's earliest contributions to physics was his study of motion. At the Cathedral of Pisa, he noticed a chandelier hanging from the ceiling took the same amount of time swinging back and forth no matter how far it swung. He began testing his theory with balls, levers, and other objects, noting detailed observations of motion. This discovery later became known as the law of the pendulum. Testing theories and conducting experiments were not common during this time. Galileo's work helped pave the way for the scientific method, a process for hypothesizing and testing ideas.

Experiments on Motion and Falling Objects

Although not confirmed, some believe that Galileo experimented with dropping objects from the Leaning Tower of Pisa in Italy. He dropped cannonballs, wood, and other objects. He would drop two of the same objects, the same size but in different weights, to test which one would fall first. The experiment proved the objects dropped at the same rate, regardless of the weight. Through experiments, he challenged the long-standing Aristotelian belief that heavier objects fall faster than lighter ones. It was during this time that he wrote *Du Motu*, or *On Motion*.

Revolutionizing Astronomy with the Telescope

In 1609, the telescope was invented in the Netherlands. Galileo heard of this invention and built his own version soon after. The original telescope only allowed for three times the magnification, while Galileo's new one displayed thirty times the magnification. Initially, Galileo sold his telescopes to Venetian merchants to help them spot ships arriving at the port.

Through his new telescope, Galileo learned that the Milky Way was made up of stars, the Sun had dark spots, and Earth's moon had craters on its surface. Galileo wrote *Siderius Nuncious*, or *The Starry Messenger*, which detailed his observations of the Moon and the stars.

Astronomical Discoveries

In 1610, he discovered four objects around Jupiter (he named the Medicean stars) which turned out to be its moons. Io, Ganymede, Europa, and Callisto were named the Galilean moons in his honor. Although everything was documented, Galileo had trouble convincing others that he had indeed discovered these new objects because fellow astronomers opposed his theories. This was the first evidence of celestial bodies orbiting an object other than Earth, challenging the geocentric model of the universe.

Galileo discovered that Venus had phases just like Earth's moon, which supported the ideas of an astronomer named Copernicus. Copernicus was the first to suggest that the Sun was the center of the solar system. This was a highly controversial topic at the time because people believed the Sun and planets revolved around the Earth. Galileo's observations provided evidence to support Copernicus' theory of the heliocentric model. In 1632, Galileo wrote a book called the *Dialogue Concerning the Two Chief World Systems* with details of his findings which supported Copernicus.

Conflict with the Roman Catholic Church

Galileo had a very difficult life. Every time he discovered something new, people did not believe him. When he claimed that everything revolved around the Sun, the Roman Catholic Church was angered. His discoveries went against everything the church believed. Galileo was put on trial for heresy (a belief that went against a religious law) in 1633. He was forced to withdraw his ideas and was not allowed to leave his home for many years as punishment. Only after his death in 1642 did the church and people begin to believe some of his theories. Today, Galileo is known as the "Father of Modern Science."

TRUE OR FALSE

After reading about **Galileo Galilei**, read each statement below and determine if it is true or false. If the statement is true, color the coin that corresponds with that question. If the statement is false, cross out that coin value. When you are finished, add the TOTAL of **ALL TRUE** coin values to reveal a 4-digit code. One digit of the code has been provided for you. If the total is 625, a 6 would go in the first box, the 2 in the second box and so on.

A. Initially, Galileo sold his telescopes to Venetian merchants to help them spot ships arriving at the port.

B. Today, Galileo is known as the "Father of our Country."

C. When Galilei attended the University of Pisa, he studied politics, hoping to become a lawyer.

D. *Du Motu* detailed Galileo's observations of the Moon and the stars.

E. Galileo discovered that Venus had phases just like Earth's moon.

F. Galileo Galilei was an astronomer, scientist, mathematician, inventor, and accomplished musician.

G. In 1608, Galileo wrote *Dialogue Concerning the Two Chief World Systems.*

H. In 1609, the telescope was invented in the Netherlands.

After shading the coins based on your answer, add the value of ALL TRUE statements to get the final total. Record your answer in the boxes below.

DOUBLE PUZZLE

After reading about **Galileo Galilei**, determine the word that corresponds with the statements provided below. Spell the corresponding word in the boxes to the right. You may or may not use all squares provided for each answer. Any numerical answers must be spelled out. Next, use the numbers **under** indicated letters to reveal a secret word.

#	Clue									
1	Galileo was the eldest of this many children									
2	Galilean moons were discovered around which planet						6			
3	A belief that went against a religious law		8							
4	Today, Galileo is known as the "Father of ___ Science"	5		7						
5	Country where Galileo was born					4				
6	Galileo wrote Siderius ___	3								
7	The four objects around Jupiter were first named the Medicean ___			2						
8	Number of Galilean moons discovered by Galileo									
9	Galileo's last name	1								
10	Galileo's telescope displayed ___ (#) times the magnification									

SECRET WORD

1	2	3	4	5	6	7	8

THE SUN AND THE MOON

THE SUN

The Sun is a giant ball of hot gas that generates energy through nuclear fusion in its core, providing the light and heat necessary to sustain life on Earth. The Sun is a yellow dwarf star (G-type main-sequence star) at the center of our solar system and is approximately 4.6 billion years old. All planets, dwarf planets, comets, and asteroids in the solar system orbit the Sun due to its immense gravitational pull.

Additionally, the entire solar system, including the Sun, orbits around the center of our galaxy, the Milky Way, in an orbit that takes roughly two-hundred-twenty-five million years to complete, known as the galactic year.

The Structure of the Sun

The Sun is so large that it makes up 99.86% of all the mass in our solar system. With seven layers in total, the Sun is composed of three inner layers and four outer layers. The inner layers include the core, the radiative zone, and the convection zone.

At the Sun's center lies the core, where temperatures soar to about 15 million degrees Celsius (28 million degrees Fahrenheit). This is where nuclear fusion occurs, producing the Sun's immense energy.

Surrounding the core is the radiative zone. Here, energy slowly moves outward in the form of radiation. It can take thousands of years for a photon of light to pass through this layer. Above the radiative zone lies the convective zone, where hot plasma rises and cooler plasma sinks, creating convection currents. These movements transport energy to the surface.

The outer layers include the photosphere, the chromosphere, the transition region, and the corona. The photosphere is the Sun's visible surface, emitting the light we see on Earth. It is relatively cool compared to the inner layers, at about 5,500 degrees Celsius (9,932 degrees Fahrenheit).

Beyond the photosphere are the chromosphere and the corona. Prominent structures in the chromosphere are jets of plasma called spicules. These plasma jets shoot upwards from the chromosphere into the lower corona. Solar spicules were first observed by Angelo Secchi, an Italian astronomer, in 1877. The corona, surprisingly hotter than the surface, extends millions of kilometers into space and is visible during a solar eclipse.

Composition, Energy and Fusion

The Sun is predominantly composed of hydrogen (about 74%) and helium (about 24%), with trace amounts of heavier elements like carbon, oxygen, and iron. At the core, hydrogen atoms fuse to form helium in a process known as nuclear fusion. This reaction releases an enormous amount of energy, as a small amount of the hydrogen's mass is converted into energy according to Einstein's equation, $E=mc^2$. In this formula, E represents energy, m is the mass, and c represents the speed of light. Astrophysicist Arthur Eddington was the first to theorize that stars generate their energy by fusing hydrogen into helium.

The energy generated in the core of the Sun is radiated outward and eventually escapes as sunlight. The average time it takes for sunlight to reach Earth is about eight minutes. This energy powers all life on Earth and drives climate and weather patterns. Every second, the Sun converts about 4.2 million metric tons of matter into energy. The journey of energy from the core to the photosphere is incredibly slow. Photons produced in the core collide with particles in the dense radiative zone, undergoing countless interactions before escaping. The total energy the Sun emits is measured as solar luminosity. Any fluctuations in the Sun's output, whether through natural solar cycles or other processes, can affect global temperatures.

Solar Cycles

The Sun experiences an eleven-year cycle of activity, known as the solar cycle. The cycle was first suggested in 1838 by astronomer Samuel Schwabe. This cycle is marked by variations in the Sun's magnetic field, which influence sunspots, solar flares, and coronal mass ejections. Sunspots are dark, cooler areas on the Sun's surface caused by magnetic activity. At the solar minimum, the Sun has the least number of sunspots and reduced solar activity. At the solar maximum, sunspot numbers peak, and solar activity, including solar flares and ejections, becomes more intense.

Solar flares are intense bursts of radiation caused by magnetic energy release, while Coronal Mass Ejections (CME) are massive expulsions of solar plasma and magnetic fields. These phenomena can disrupt satellites, power grids, and communication systems on Earth.

The Sun's Role

The Sun's gravity holds the planets, comets, asteroids, and other celestial objects in orbit. Its energy drives photosynthesis on Earth, powers the water cycle, and influences planetary climates. The Sun's radiation also generates the solar wind, a stream of charged particles that shapes the heliosphere. The heliosphere is the vast bubble of space dominated by the Sun's influence.

PARAGRAPH CODE

After reading about the **Sun**, head back to the reading and number ALL the paragraphs in the reading passage. Then, read each statement below and determine which paragraph **NUMBER** the statement can be found in. Paragraph numbers MAY be used more than one time or not at all. Follow the directions below to reveal the 4-digit code.

A The outer layers include the photosphere, the chromosphere, the transition region, and the corona.

B The heliosphere is the vast bubble of space dominated by the Sun's influence.

C These plasma jets shoot upwards from the chromosphere into the lower corona.

D Astrophysicist Arthur Eddington was the first to theorize that stars generate their energy by fusing hydrogen into helium.

E The total energy the Sun emits is measured as solar luminosity.

F The Sun experiences an eleven-year cycle of activity, known as the solar cycle.

G The Sun is so large that it makes up 99.86% of all the mass in our solar system.

H The Sun is a yellow dwarf star (G-type main-sequence star) at the center of our solar system.

ELIMINATE ALL EVEN-NUMBERED paragraphs that you used as an answer. Record the remaining numbers (in the SAME order in which you recorded them above) in the boxes below.

MYSTERY WORD

After reading about the **Sun**, determine if each statement below is true or false. Color or shade the boxes of the **TRUE** statements. Next, unscramble the mystery word using the large letters of the **TRUE** statements.

Surrounding the core is the radiative zone. **A**	The average time it takes for sunlight to reach Earth is about forty minutes. **S**	The Sun is predominantly composed of hydrogen and helium. **R**	Solar spicules were first observed by Albert Einstein. **L**
The solar cycle was first suggested in 1838 by astronomer Samuel Schwabe. **P**	The total energy the Sun emits is measured as solar luminosity. **E**	The Sun is so large that it makes up 99.86% of all the mass in our solar system. **T**	Photons produced in the core collide with particles in the dense radiative Zone. **R**
The Sun is a green dwarf star (Y-type main-sequence star). **B**	The Sun is composed of five inner layers and two outer layers. **C**	The Sun is a giant ball of hot gas. **E**	Every second, the Sun converts about 4.2 million metric tons of matter into energy. **M**
The Sun experiences a three-year cycle of activity, known as the solar cycle. **D**	The photosphere is the Sun's visible surface, emitting the light we see on Earth. **U**	Solar flares are intense bursts of radiation caused by magnetic energy release. **E**	Sunspots are dark, cooler areas on the Sun's surface caused by magnetic activity. **T**

Unscramble the word using the large bold letters of <u>only</u> the **TRUE** statements.

PHASES OF THE MOON

The Moon is the Earth's largest satellite and fifth largest moon in the solar system. A satellite is a celestial body that orbits a planet. As the Moon travels around the Earth it forms an ellipse, not a perfect circle. Because of its orbit, the Moon goes through different phases each month. A phase is how the Moon looks when viewed from Earth.

Many ancient cultures, including the Babylonians, Greeks, and Chinese, developed lunar calendars based on the phases of the Moon. In Greek mythology, the Moon goddess was known as Selene. A Mesolithic, or Middle Stone Aged, arrangement of twelve pits that form an arc was found in Scotland. The pits date back to 8,000 BC and became known as the "world's oldest known calendar" in 2013.

The Moon is about a quarter the size of the Earth. Due to gravity, something on the Moon will be six times lighter than on the Earth. The nightly temperature on the Moon is -280 degrees Fahrenheit.

The Moon's outermost layer is its crust, which is primarily composed of oxygen, silicon, magnesium, iron, calcium, and aluminum. The most abundant rock types found in the Moon's crust are basalts and anorthosites. Basalts form from volcanic activity, and anorthosites are composed mostly of a mineral called plagioclase feldspar.

The surface of the Moon is covered with a layer of loose, fragmented material called regolith. This layer is a mix of dust and small rocks.

The Moon's Orbit

As the Moon orbits the Earth, the phases of the Moon change. The part of the Moon facing the Sun will reflect its light and become visible. If the Sun did not exist, then we would not be able to see the Moon. Known as albedo, the Moon only reflects seven percent of sunlight. When the Moon appears to be getting bigger, it is called waxing. When it looks as though it is getting smaller, it is called waning.

The time it takes the Moon to make a complete cycle of phases is called a lunar month. A lunar month (29.53 days) is shorter than a standard calendar month (30.44 days). During the first few phases, the Moon is waxing and becoming more visible. In the second half, the Moon is waning until it cannot be seen any longer.

Gravity and Tides

The gravitational pull of the Moon affects Earth's oceans, causing high and low tides. The highest tides occur during the Full Moon and New Moon when the Sun, Earth, and Moon are aligned. The Moon's rotation is synchronous with its orbit around Earth, which means that it takes the same amount of time for the Moon to rotate once on its axis as it does to orbit Earth. This is why we always see the same side of the Moon from Earth.

Moon Phases

The first phase cannot be seen from Earth and is known as the New Moon (also called a Dark Moon). This only happens when the Moon is directly between the Sun and the Earth. Both the Sun and the Moon have the same ecliptic longitude. The Sun is shining only on the part of the Moon that is facing away from Earth, and the part of the Moon facing the planet is completely dark.

Three or four nights after a New Moon will be a Waxing Crescent, when the Moon is seen as a small crescent shape. It becomes more visible each day as the Sun's light reflects more of its surface.

The next phase is the First Quarter, or a Half Moon. During this phase, exactly half (50%) of the Moon is illuminated by sunlight because it has completed the first quarter of its orbit.

The fourth phase, Waxing Gibbous, is when more than half the Moon is visible. The Moon is heading toward the East by about 12 degrees per day, away from the Sun. As the Moon moves away, it waxes or appears larger.

During a Full Moon, the entire half of the Moon that faces the Earth can be seen as a circle. This phase happens when the Moon is on the opposite side of the Earth from the Sun. The time between one Full Moon and the next is 29.5 days. Therefore, it is possible for a month to have two Full Moons. There would be one at the very beginning of the month and one at the end of the month. This occurs every few years and the second Full Moon is called a "Blue Moon." Because February is a short month, sometimes there is no Full Moon phase.

Following the Full Moon is the Waning Gibbous. More than half of the Moon is visible, but each day it is waning or becoming less visible. On average, the Moon rises approximately 50 minutes later each night. It continues to get smaller until it becomes the Last Quarter Moon.

Finally, the Waning Crescent is the last phase. It is a crescent shaped Moon that becomes smaller each day until it is no longer visible. This phase occurs when the Moon is on the opposite side of the Earth from the Sun. The cycle then starts again with the New Moon.

MYSTERY MATCH

After reading about the **Phases of the Moon**, draw a line from the left-hand column to make a match in the right-hand column. Your line should go through **ONE** letter. When you complete all the matches, use the letters with a line THROUGH them to unscramble a mystery word. You MUST start and end your line at the **arrow points**.

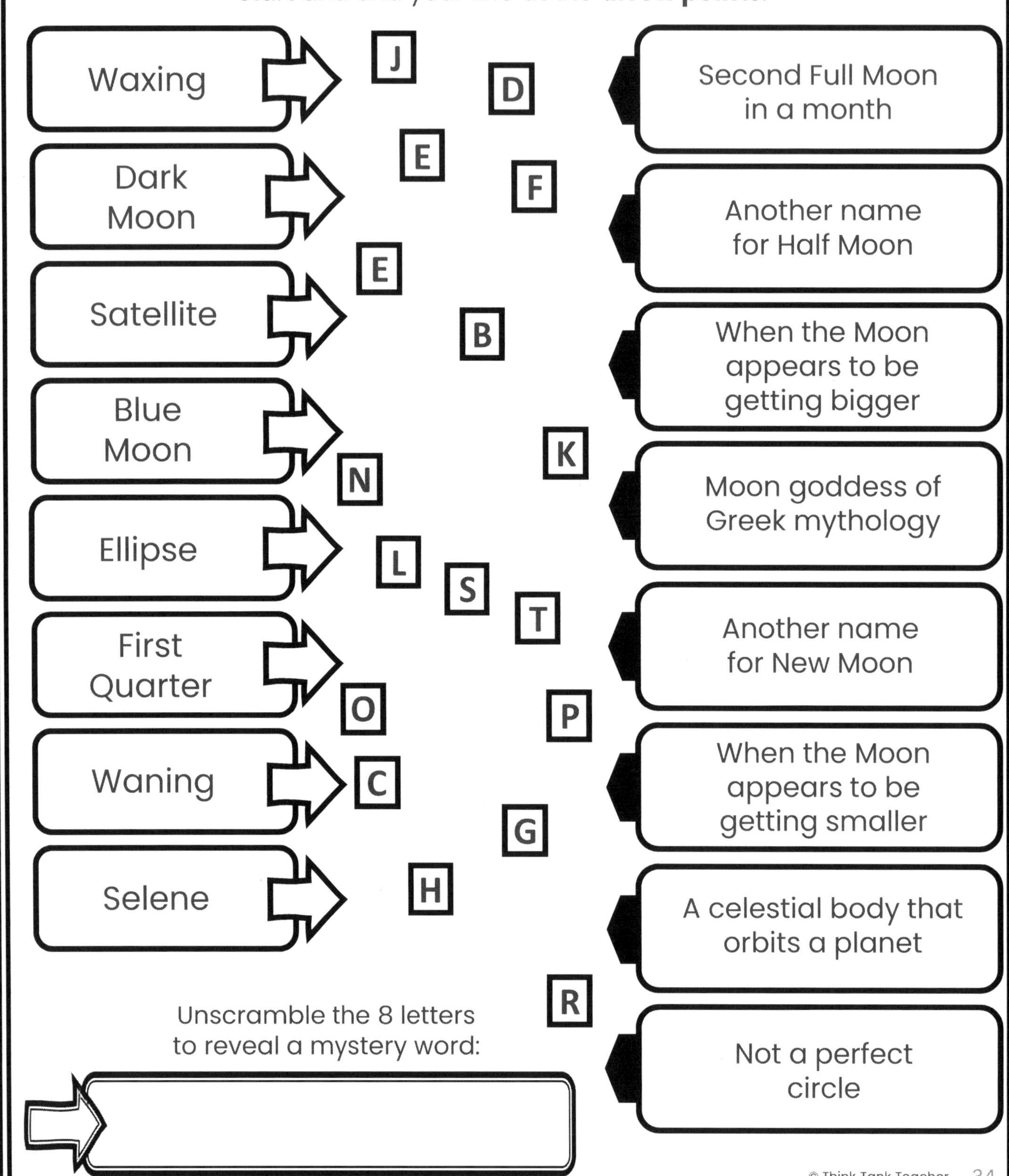

Unscramble the 8 letters to reveal a mystery word:

MULTIPLE CHOICE

After reading about the **Phases of the Moon**, answer each multiple-choice question below. Then, count the number of times you used each letter as an answer (ABCD) to reveal a 4-digit code. Letters may be used more than once or not at all. If a letter option is not used, put a zero in that box.

1 What percent of the Moon is visible during the First Quarter phase?

A. 10%
B. 25%
C. 50%
D. 65%

2 In terms of moons in our solar system, where does Earth's moon rank?

A. Second largest moon
B. Third largest moon
C. Fourth largest moon
D. Fifth largest moon

3 Due to gravity, something on the Moon will be how many times lighter than on the Earth?

A. Six
B. Twelve
C. Eighteen
D. Thirty

4 What is the Moon's outermost layer composed of?

A. Iron
B. Calcium
C. Magnesium
D. All of the above

5 What is the time it takes the Moon to make a complete cycle of phases known as?

A. Lunar month
B. Lunar albedo
C. Lunar day
D. None of the above

6 Which phase occurs three or four nights after a New Moon?

A. Third Quarter
B. Waning Gibbous
C. Waxing Crescent
D. Blue Moon

7 Which ancient culture developed lunar calendars based on the phases of the Moon?

A. Chinese
B. Greeks
C. Babylonians
D. All of the above

8 How often does a Full Moon occur?

A. About every 10 days
B. About every 20 days
C. About every 30 days
D. About every 40 days

Count how many times you used each letter as a correct answer (ABCD) to determine the 4-digit code. Record your answer in the boxes below.

# of A's	# of B's	# of C's	# of D's

INNER PLANETS

MERCURY

Mercury is the smallest of the eight planets in the solar system and the one closest to the Sun. Still, Venus is a hotter planet than Mercury because of its dense atmosphere. Like all objects in the solar system, it revolves around the Sun. It is gray in color and has no known moons or rings. The first four planets (Mercury, Venus, Earth, Mars) are known as the inner rocky planets, or terrestrial planets.

Historical Observations

The first known record of Mercury was 4th century BC by Assyrian astronomers. Experts believe that the Sumerians and Babylonians from around 3000 BC knew about Mercury. Yet, Galileo was the first to see Mercury through a telescope in the 1600s. Scientists believe the earliest human civilizations knew about Mercury because it can be seen in the sky on a dark night. The planet was named after the Roman god Mercury who was the god of travel and thievery. To ancient Greeks, the planet was known as Hermes.

Temperature and Physical Characteristics

The time it takes for sunlight to reach Mercury is about three minutes. Temperatures on this planet can reach up to 800 degrees Fahrenheit during the day and as low as -300 degrees Fahrenheit during the night when it is facing away from the Sun. Mercury does not have an atmosphere like Earth. Instead, there is only an exosphere with oxygen, sodium, helium, hydrogen, and potassium. Mercury does not have seasons. This is because the tilt in the axis of rotation is only two degrees. Mercury spins upright, unlike Earth.

Mercury has a diameter of 3,031 miles and a radius of 1,516 miles. The planet is solid, composed of rocks. In the center of the planet, there is a large iron core. The core makes up about 85% of the planet's volume. The core's significant size relative to the planet suggests that Mercury underwent a unique formation process or experienced significant stripping of its outer layers, possibly early in its history.

Similar to Earth's Moon, the surface of Mercury is covered with craters of all sizes. The craters came about over billions of years because of comets and asteroids that have hit the planet. One of the largest craters, at about 950 miles across, is called the Caloris Basin. It was a result of an asteroid that scientists believe was 60 miles wide. The large crater is big enough for the state of Texas to fit inside.

Interestingly, many craters on Mercury are named after famous authors, musicians, and artists. For example, one crater is named after Dr. Seuss, the famous children's author, and another is named after Alvin Ailey, a famous dancer. Maya Angelou, Ernest Hemingway, Mark Twain, Edgar Allan Poe, Walt Disney, Andy Warhol, and Walt Whitman also have a crater named in their honor. There are over four hundred named Mercurian craters. The tradition of naming craters on Mercury after famous figures began in 1976. The names are selected by the International Astronomical Union (IAU).

Orbital and Rotational Dynamics

Time on Mercury is different than on Earth. While a year on Earth takes 365 days, it only takes 88 days for Mercury to revolve around the Sun. It is one of the fastest planets in the Sun's orbit and moves very quickly at 50 km per second. However, it rotates much slower than Earth and one rotation, or day, on Mercury takes 59 Earth days. The Sun does not rise and set on the planet as it does on Earth. Because of the quick revolution and slow rotation, it takes 180 Earth days for the Sun to rise on Mercury. Due to the lack of atmosphere, Mercury does not experience weather, clouds, wind, or storms.

Mercury's pathway around the Sun is an odd elliptical shape, like an egg. Known as retrograde, Mercury appears to travel backward in the sky three to four times per year. This phenomenon occurs due to the shorter orbit of the planet, when it is moving slower than Earth around the Sun.

Challenges of Exploration

Because Mercury is the closest planet to the Sun, it is difficult for scientists to learn more about the planet. The extreme temperatures and strong gravitational force from the Sun make it impossible for astronauts to visit the planet. Sending a spacecraft is also difficult. The first spacecraft was the Mariner 10 which launched in the 1970s and passed by Mercury three times. It took many photos that showed what the planet looked like.

The Messenger was the second spacecraft, launched in 2004. Seven years later in 2011, it began to orbit the planet which lasted for four years. The entire mission lasted a total of 10 years and 8 months before it crashed into Mercury in 2015. The craft studied the planet and provided information to scientists that would never have been known otherwise such as surface composition, the geological history, and its magnetic field.

People often wonder if life could exist on other planets. However, with the extreme temperatures and solar radiation, organisms could not adapt and survive on Mercury.

TRUE OR FALSE

After reading about **Mercury**, read each statement below and determine if it is true or false. If the statement is true, color the coin that corresponds with that question. If the statement is false, cross out that coin value. When you are finished, add the TOTAL of **ALL TRUE** coin values to reveal a 4-digit code. One digit of the code has been provided for you. If the total is 625, a 6 would go in the first box, the 2 in the second box and so on.

A. One of the largest craters, at about 950 miles across, is called the Caloris Basin.

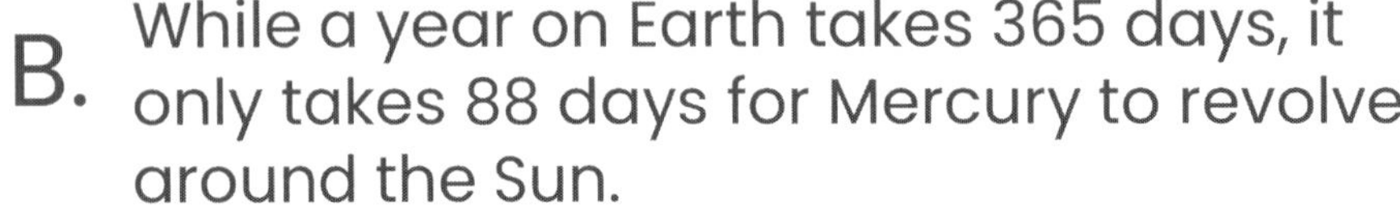

B. While a year on Earth takes 365 days, it only takes 88 days for Mercury to revolve around the Sun.

B
25

C. Mercury is the largest of the eight planets in the solar system.

D. Due to the lack of atmosphere, Mercury does not experience weather, clouds, wind, or storms.

C
50

E. The time it takes for sunlight to reach Mercury is about three minutes.

F. The first four planets (Mercury, Venus, Earth, Mars) are known as the inner rocky planets, or terrestrial planets.

D
100

G. The tradition of naming craters on Mercury after famous figures began in 1976.

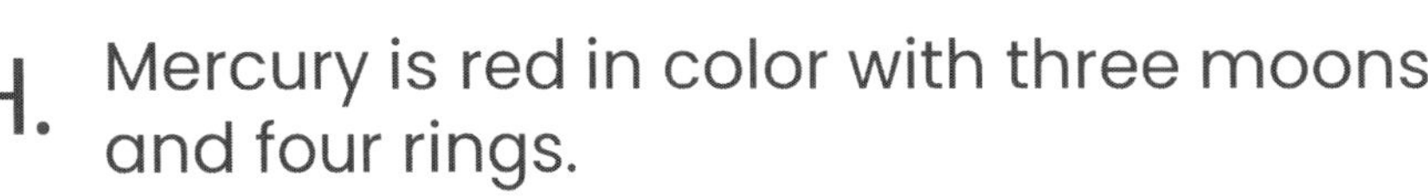

H. Mercury is red in color with three moons and four rings.

After shading the coins based on your answer, add the value of ALL TRUE statements to get the final total. Record your answer in the boxes below.

DOUBLE PUZZLE

After reading about **Mercury,** determine the word that corresponds with the statements provided below. Spell the corresponding word in the boxes to the right. You may or may not use all squares provided for each answer. Any numerical answers must be spelled out. Next, use the numbers **under** indicated letters to reveal a secret word.

1	Terrestrial planets are also known as the ___ rocky planets				5					
2	Known as retrograde, Mercury appears to travel ___			1						
3	To ancient Greeks, the planet was known as this						6			
4	The first known record of Mercury was 4th century BC by ___ astronomers		4							
5	There are over ___ hundred named Mercurian craters									
6	At 950 miles across, the Caloris Basin was a result of an ___			7						
7	International Astronomical ___ (IAU)									
8	He was the first to see Mercury through a telescope			2						
9	In the center of Mercury, there is a large ___ core			3						
10	The first spacecraft to visit Mercury was the ___ 10									

SECRET WORD

1	2	3	4	5	6	7

VENUS

Venus is the sixth largest planet in the solar system, closest planet to Earth, and the second closest planet to the Sun. It shines brighter than any other planet seen on Earth, and brighter than anything in the night sky besides the Moon. Its brightness, caused by its reflective clouds, makes it visible without a telescope.

Although the exact time of discovery is unknown, early civilizations observed the planet thousands of years ago since they could easily spot it in the sky. Ancient civilizations such as the Babylonians, Greeks, Egyptians, and Mayans meticulously recorded the movements of Venus.

Initially, people thought Venus was two separate planets or stars: the morning star and the evening star. The morning star was called Phosphorus ("light-bringer" in Greek) because dawn would come soon in the east after it appeared. The evening star was called Hesperus ("west" in Greek) because it appeared in the west after sunset. However, Pythagoras, a Greek mathematician, thought it was one planet. So, the Greeks named the planet Aphrodite after the Greek goddess of love and beauty. Romans changed the name to Venus. In 1610, Galileo Galilei observed Venus through a telescope and noted its phases, confirming it orbited the Sun.

The Atmosphere

Venus has a very thick atmosphere made of mostly carbon dioxide, nitrogen, and clouds of sulfuric acid (a poisonous gas to humans). Because more than 96% of the atmosphere is carbon dioxide, it traps a large amount of heat from the Sun, making it the hottest planet in the solar system. Venus' carbon dioxide atmosphere is ninety times as thick as Earth's. On Venus, temperatures can reach over 800 degrees Fahrenheit. The heat is so intense that metals on the planet, such as lead, melt. To reach Venus, the light from the Sun takes about six minutes.

Like Mercury, there are no moons that orbit Venus. Even though there are clouds, it is too hot to rain on Venus. The clouds travel nearly three times faster than hurricane winds on Earth, making a path around Venus every four days. A thunder and lightning show can be seen beneath the layer of clouds, but no rain. The thick clouds of sulfuric acid reflect about 75% of the sunlight that reaches the planet, giving it a bright appearance.

Venus's thick atmosphere makes it impossible to see the surface even in space. Despite being closer to the Sun, about forty spacecraft have visited Venus. What scientists have learned about the planet has been through ultraviolet cameras and radars.

The Rocky Surface

Venus has a rocky composition with a central core, a molten mantle, and a solid crust. The mantle is a thick, silicate layer rich in magnesium and iron. Venus is a terrestrial planet meaning it is solid and made of rocks. The high temperatures make it a desert with no water. The surface is covered with thousands of mountains and volcanoes which are believed to have been active billions of years ago. Venus has more volcanoes than any other planet in the solar system. Scientists believe some volcanoes may still be active today.

Some of the craters on Venus look like spiders, so they are called "arachnids." Maxwell Montes is the highest mountain on Venus, similar in size to Mount Everest. Sif Mons is one of the largest volcanoes on Venus. It was named after the Norse goddess Sif, with the name officially adopted in 1982. The highland region is known as Ishtar Terra.

Unlike Earth and most of the planets, Venus rotates (spins) in the opposite direction. This means the Sun rises in the west and sets in the east. It also rotates very slowly, taking 243 Earth days to make one rotation. A revolution, however, is shorter, taking only 225 Earth days. This is due to the fact that it is closer to the Sun and takes less time to go around. Venus only has two sunrises during its year because the length of the day and year are very similar.

Space Exploration of Venus

Americans were the first on the Moon, but Russians were the first to land on Venus with the Venera 7 spacecraft. The first successful mission to Venus took place on December 14, 1962, however, it only lasted 23 minutes before it was destroyed by the conditions. From 1989 to 1994, the Magellan Probe, launched by the United States, orbited the planet and was able to map the surface in detail. The Magellan was able to map 98% of the planet's surface. In 2006, the European Space Agency sent the Venus Express to gather information. Other successful launches and missions have allowed scientists to learn about the planet, its atmosphere, and surface.

Sometimes, Venus is considered Earth's sister because its size, mass, and density are similar. Venus' mass is about 82 percent of Earth's mass. The diameter of Venus is 7520 miles (12,104 km). Experts think that Venus had water and oceans, like Earth, at one point, too. It has since boiled away.

PARAGRAPH CODE

After reading about **Venus**, head back to the reading and number ALL the paragraphs in the reading passage. Then, read each statement below and determine which paragraph **NUMBER** the statement can be found in. Paragraph numbers MAY be used more than one time or not at all. Follow the directions below to reveal the 4-digit code.

A Americans were the first on the Moon, but Russians were the first to land on Venus with the Venera 7 spacecraft.

B Venus has more volcanoes than any other planet in the solar system.

C Ancient civilizations such as the Babylonians, Greeks, Egyptians, and Mayans meticulously recorded the movements of Venus.

D Initially, people thought Venus was two separate planets or stars: the morning star and the evening star.

E Venus is the sixth largest planet in the solar system, closest planet to Earth, and the second closest planet to the Sun.

F Venus has a very thick atmosphere made of mostly carbon dioxide, nitrogen, and clouds of sulfuric acid (a poisonous gas to humans).

G Maxwell Montes is the highest mountain on Venus, similar in size to Mount Everest.

H Venus has a rocky composition with a central core, a molten mantle, and a solid crust.

ELIMINATE ALL EVEN-NUMBERED paragraphs that you used as an answer. Record the remaining numbers (in the SAME order in which you recorded them above) in the boxes below.

MYSTERY WORD

After reading about **Venus**, determine if each statement below is true or false. Color or shade the boxes of the **TRUE** statements. Next, unscramble the mystery word using the large letters of the **TRUE** statements.

The mantle is a thick, silicate layer rich in magnesium and iron. **T**	Germans were the first to land on Venus with the Magellan 7 spacecraft. **D**	Unlike Mercury, seven moons orbit Venus. **P**	Venus is the fourth closest planet to the Sun. **F**
The highland region is known as Ishtar Terra. **I**	The morning star was called Phosphorus ("light-bringer" in Greek). **H**	To reach Venus, the light from the Sun takes about six minutes. **E**	Venus' mass is only about 16 percent of Earth's mass. **L**
Greeks named the planet Aphrodite after the Greek goddess of love and beauty. **R**	Olympus Mons is the highest mountain on Venus. **N**	Even though there are clouds, it is too hot to rain on Venus. **S**	Venus takes 243 Earth days to make one rotation. **G**
The first successful mission to Venus took place on June 4, 1943. **A**	Venus only has two sunrises during its year. **T**	Some of the craters on Venus look like spiders, so they are called "arachnids." **B**	Venus is the third largest planet in the solar system. **C**

Unscramble the word using the large bold letters of only the **TRUE** statements.

EARTH

Earth is the third planet from the Sun and the fifth largest planet in the solar system. It is inhabited by humans and living organisms and is the only planet with oxygen, which makes it possible to support life. The Sun provides the Earth with energy and heat that humans, plants, and animals need to survive.

The Earth has one moon in its orbit. Although the Moon looks smooth when viewed from Earth, in reality, it is covered with craters (holes). These craters were formed by impact such as meteorites.

The word "Earth" comes from an Anglo-Saxon word "Erda," which means soil or ground. Of all the planets in the solar system, Earth is the only planet that was not named after a god or goddess.

Earth's Movement: Rotation and Revolution

The Earth, commonly referred to as the Blue Planet, revolves around the Sun every 365.25 days. This revolution is known as a year. The Earth also rotates and takes 24 hours to make one rotation. The side facing the Sun has daytime while the side that faces away experiences nighttime. A year on Earth is 365 days long but the ¼ days are added up. Every fourth year has one extra day. The extra day is February 29th. This fourth year is called a Leap Year and has 366 days.

Earth spins on an imaginary line from the North pole to the South pole, called the axis. Due to the tilt of Earth in relation to the Sun, Earth experiences an equinox twice a year. During the Autumnal and Vernal equinox, the amount of day and night is about equal. Most places on Earth experience four seasons, caused by the Earth's tilt and rotation.

Earth's Atmosphere

Earth's atmosphere consists of five main layers: the troposphere, stratosphere, mesosphere, thermosphere, and the exosphere. These layers of gases work together to keep the Earth warm, provide oxygen that all living things need, and is where the weather takes place. The troposphere is closest to the surface where humans, plants, and animals live. The stratosphere is above the troposphere and contains the ozone layer. This important layer protects the Earth from the Sun's dangerous radiation. The mesosphere protects the Earth from meteors and is the coldest region. The next layer is the thermosphere which is where the International Space Station and other Earth satellites are in orbit. Finally, the exosphere is the outermost layer closest to outer space.

Earth's Structure

The Earth is one of the four inner terrestrial planets, meaning it has a hard rocky surface and iron core. Earth is divided into three main layers: the crust, the mantle, and the core.

The core is the layer in the middle of the Earth which reaches 10,800 degrees Fahrenheit. It is not possible for humans to reach the core. The core is further divided into the inner core and the outer core. The outer core is a liquid layer composed mainly of iron and nickel. The movement of the molten iron in the outer core creates electric currents, which in turn produce the planet's magnetic field, protecting Earth from solar radiation and helping to maintain life. The inner core has a radius of about 760 miles. The inner core is under immense pressure, causing it to be solid despite the high temperatures. Its solid state and heat contribute to the convection currents in the outer core.

The mantle is the middle layer. It is a thick layer that extends to about 1,800 miles below the surface. The mantle makes up over 80% of Earth's mass (weight). The lithosphere is where the crust and mantle meet and where tectonic plates can be found. These plates are constantly moving at a very slow pace.

The crust is the outer layer where all life exists. The surface of the Earth is covered with 71% water and 96.5% of that is made up of the oceans. Earth has changed a lot since it was first formed billions of years ago. Collisions and earthquakes have had an impact on how the planet looks today.

The Formation of Continents

The land that covers the Earth was once one large piece named Pangea. Billions of years ago, it broke apart because of the movements of the tectonic plates in the lithosphere and eventually became the seven continents. The seven continents include Asia, Africa, North America, South America, Australia, Europe, and Antarctica. Earth has five major oceans which include the Atlantic, Pacific, Indian, Antarctic (Southern) and Arctic Oceans.

Mount Everest is the highest point above sea level. The tallest mountain from the base to the summit is Mauna Kea in Hawaii, however, most of this mountain is under water. The lowest point is the Mariana Trench in the Pacific Ocean. The Nile River is the longest river on Earth at 4160 miles long. The Atacama Desert in South America is the driest place on Earth.

MYSTERY MATCH

After reading about **Earth**, draw a line from the left-hand column to make a match in the right-hand column. Your line should go through **ONE** letter. When you complete all the matches, use the letters with a line THROUGH them to unscramble a mystery word. You MUST start and end your line at the **arrow points**.

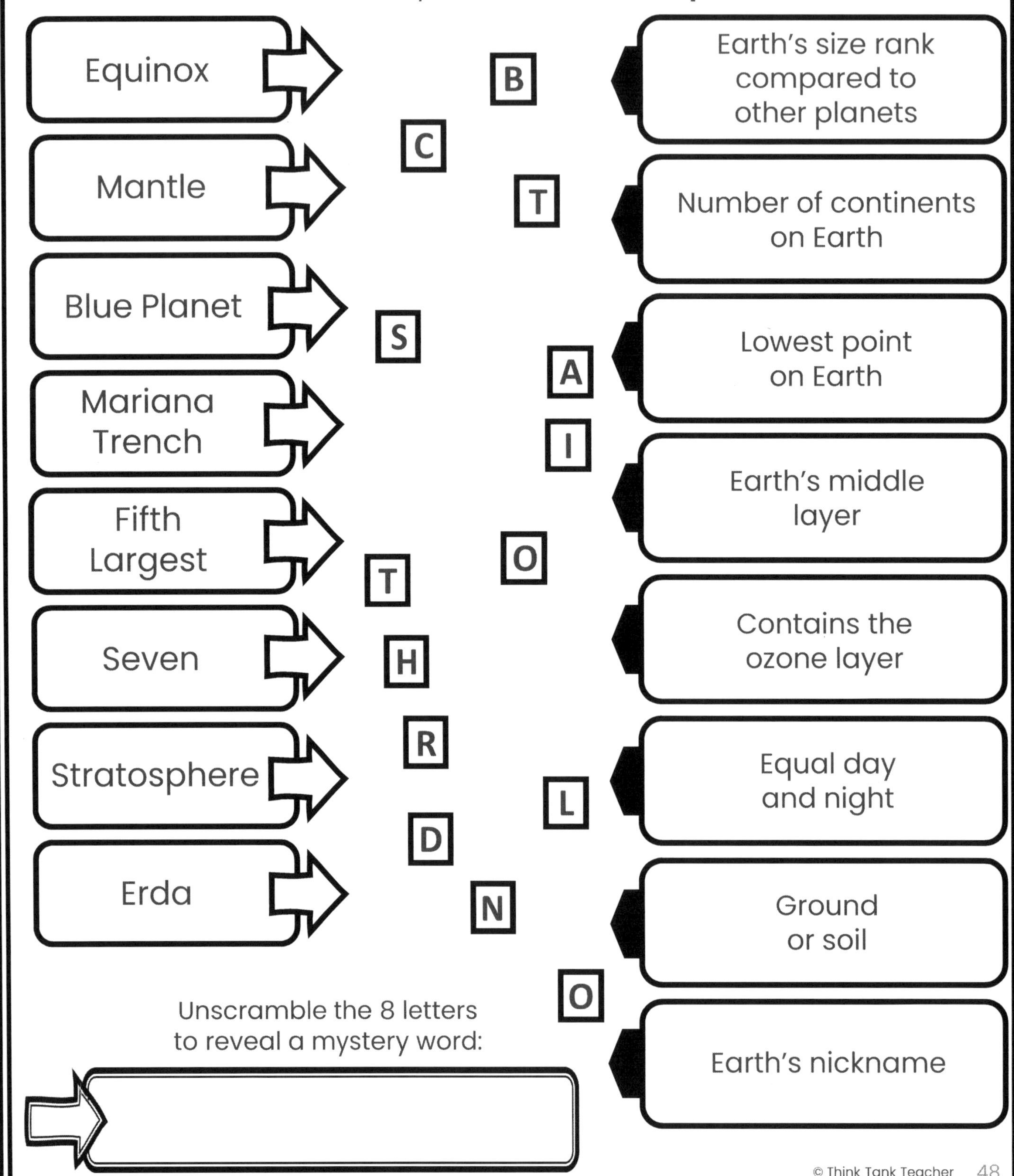

Equinox
Mantle
Blue Planet
Mariana Trench
Fifth Largest
Seven
Stratosphere
Erda
B
C
T
S
A
I
O
T
H
R
L
D
N
O
Earth's size rank compared to other planets
Number of continents on Earth
Lowest point on Earth
Earth's middle layer
Contains the ozone layer
Equal day and night
Ground or soil
Earth's nickname

Unscramble the 8 letters to reveal a mystery word:

MULTIPLE CHOICE

After reading about **Earth**, answer each multiple-choice question below. Then, count the number of times you used each letter as an answer (ABCD) to reveal a 4-digit code. Letters may be used more than once or not at all. If a letter option is not used, put a zero in that box.

1 What is the longest river on Earth?

A. Rio Grande River
B. Thames River
C. Nile River
D. Amazon River

2 In the solar system, how many inner terrestrial planets are there?

A. Two
B. Four
C. Six
D. Eight

3 What imaginary line from the North pole to the South pole does the Earth spin on?

A. Axis
B. Equator
C. Tropic of Cancer
D. None of the above

4 What is the outer core mainly composed of?

A. Helium and neon
B. Iodine and basalt
C. Neon and sulfur
D. Iron and nickel

5 How many main layers does Earth's atmosphere consists of?

A. Three
B. Five
C. Seven
D. Nine

6 Which ocean is also known as the Southern Ocean?

A. Antarctic
B. Pacific
C. Atlantic
D. Arctic

7 What percent of Earth's mass (weight) does the mantle make up?

A. 80%
B. 60%
C. 40%
D. 20%

8 How many moons does the Earth have in its orbit?

A. One
B. Two
C. Three
D. Four

Count how many times you used each letter as a correct answer (ABCD) to determine the 4-digit code. Record your answer in the boxes below.

# of A's	# of B's	# of C's	# of D's

MARS

Mars is the fourth planet from the Sun in the solar system. It is known as the Red Planet because iron minerals found in the soil and atmosphere make it appear red. The color comes from rusting rocks (oxidation). This reddish Martian soil is called regolith. The planet looks red during a storm when the dust releases into the atmosphere. Mars is the planet where humans have searched for life the most.

The planet was formed about 4.5 billion years ago when gravity caused gas and dust to swirl and collect, forming a planet. Mars can be easily seen from Earth without a telescope and has been observed since ancient times. The earliest recorded observations can be traced to the Babylonians around 2,000 BC, who referred to the planet as "Nergal," the god of war. They meticulously recorded its movements, contributing to the foundation of astronomy. Galileo first observed Mars through a telescope in 1610.

Structure and Composition

Mars is the fourth terrestrial planet, along with Mercury, Venus, and Earth. It has a dense central core of iron, magnesium, aluminum, calcium, and potassium. Like Earth, Mars has a rocky mantle and a solid crust. The surface of Mars is dry and full of rocks.

Mars is the second smallest planet in the solar system and is about half the size of Earth. It has a lower mass and weighs about ten times less than Earth. Because of its lower mass, gravity on Mars is not as strong as it is on Earth. On Mars, a 100-pound person would weigh just 38 pounds.

Mars has an axial tilt of about 25 degrees, which is comparable to Earth's tilt of 23.5 degrees. One rotation or day on Mars takes 24.6 hours, almost as long as an Earth day. A revolution around the Sun, or a year, takes 687 days which equals almost two Earth years. This is because it is farther away from the Sun than Earth and takes longer to go around it.

Seasons and Atmosphere

Mars has seasons; however, they are much longer than on Earth. Mars' egg-shaped orbit accounts for the difference in season lengths. Spring is the longest season at 194 days, and Fall is the shortest with 142 days. Depending on the time of year and location, temperatures on Mars can range from -220 degrees Fahrenheit to 70 degrees Fahrenheit, with the average temperature of -70 degrees Fahrenheit. It takes sunlight approximately thirteen minutes to reach Mars. Mars has a very thin atmosphere made mostly of the gases carbon dioxide, nitrogen, and argon. This thin atmosphere contributes to extreme temperature fluctuations.

Moons: Phobos and Deimos

Like Earth, Mars has a moon. Actually, it has two moons shaped like potatoes, called Phobos and Deimos, discovered in 1877. Phobos goes around Mars three times in one Martian day. Deimos is one of the smallest moons in the solar system and takes a little more than a day to go around the planet. Experts think that these moons may have once been asteroids. Phobos is closest to Mars and has deep grooves. Some experts think that Deimos could crash into Mars in fifty million years.

Landmarks on Mars

On the surface of Mars, inactive volcanoes and mountains can be found. The highest mountain in the solar system, named Olympus Mons, is found on Mars, and reaches almost fifteen miles high. Olympus Mons is three times taller than Earth's Mt. Everest. It is a shield volcano, formed through tectonic phenomena similar to Earth. Valles Marineris, also on Mars, is the biggest crater in the solar system, reaching four miles deep and thousands of miles across. Compared to Earth, Valles Marineris is almost ten times bigger than the Grand Canyon.

Mars exploration

Mars is similar to Earth in many ways. It has weather, changing seasons, volcanoes, canyons, and polar ice caps. For these reasons, scientists have been attempting to learn if there has been life on Mars in the past. They have also been exploring the idea of whether life can survive there now or in the future. Today, Mars exploration is a big part of NASA's work (National Aeronautics and Space Administration).

Beginning in the 1960's, robotic spacecraft began exploring Mars. The Mariner 4 was launched by the United States in 1964. The Mariner 6 was launched by the United States in 1969. Later, the Soviet Union launched numerous missions such as the Mars 2 and Mars 3 in 1971, most of which resulted in failures. NASA'S first successful touch down on the surface of Mars was with the Viking I lander. Viking landers were the first to conduct experiments directly on the Martian surface, while InSight has been studying seismic activity to learn about Mars' internal structure. After years of finding no trace of water on the planet, evidence of a lake was found under the southern polar ice cap in 2018.

TRUE OR FALSE

After reading about **Mars**, read each statement below and determine if it is true or false. If the statement is true, color the coin that corresponds with that question. If the statement is false, cross out that coin value. When you are finished, add the TOTAL of **ALL TRUE** coin values to reveal a 4-digit code. One digit of the code has been provided for you. If the total is 625, a 6 would go in the first box, the 2 in the second box and so on.

A. Mars is the second largest planet in the solar system.

B. Deimos is one of the smallest moons in the solar system.

C. Mars is the planet where humans have searched for life the most.

D. Olympus Mons is nine times taller than Earth's Mt. Everest.

E. The Mariner 4 was launched by the Soviet Union in 1964.

F. Mars' egg-shaped orbit accounts for the difference in season lengths.

G. Mars looks red during a storm when the dust releases into the atmosphere.

H. Isaac Newton first observed Mars through a telescope in 1604.

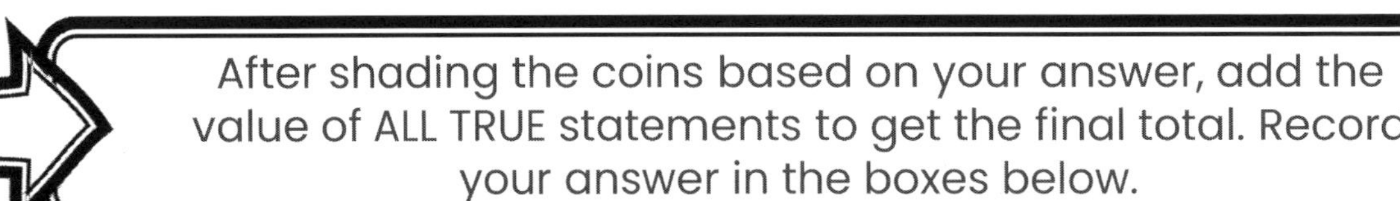

After shading the coins based on your answer, add the value of ALL TRUE statements to get the final total. Record your answer in the boxes below.

DOUBLE PUZZLE

After reading about **Mars**, determine the word that corresponds with the statements provided below. Spell the corresponding word in the boxes to the right. You may or may not use all squares provided for each answer. Any numerical answers must be spelled out. Next, use the numbers **under** indicated letters to reveal a secret word.

#	Statement	Boxes (numbered letter)
1	The longest season on Mars	7 (box 4)
2	The moon of Mars that is closest to the planet	5 (box 3)
3	___ Marineris is the biggest crater in the solar system	4 (box 3)
4	Mars has an ___ tilt of about 25 degrees	2 (box 2)
5	Phobos goes around Mars ___ times in one Martian day	1 (box 4)
6	Babylonians referred to the planet as this name for the god of war	8 (box 1)
7	Reddish Martian soil	
8	NASA'S first successful touch down on the surface of Mars was with the ___ I lander	9 (box 6)
9	The highest mountain in the solar system is ___ Mons	3 (box 5)
10	Number of minutes it takes sunlight to reach Mars	6 (box 4)

SECRET WORD

1 2 3 4 5 6 7 8 9

OUTER PLANETS

JUPITER

Did you know that Jupiter shrinks two centimeters every year because it radiates intense heat? Jupiter is the largest planet in the solar system. It is the fifth planet from the Sun and revolves around it just like all other objects in the solar system.

Jupiter can be seen from Earth without a telescope and has been known since ancient times. Jupiter was named after the king of the Roman gods in ancient mythology. The Romans named the planet after Jupiter, their god of the sky and thunder, because of its immense size and prominence in the night sky. The Greeks associated the planet with their god Zeus, who was the equivalent of Jupiter in their mythology. The Babylonians, known for their detailed astronomical observations, also recognized Jupiter as a significant celestial body, linking it to their god Marduk, the ruler of their pantheon.

Jupiter has a diameter of 88,846 miles. Jupiter's mass is over 318 times bigger than Earth. You could fit eleven Earths side by side within Jupiter's diameter. This immense size and mass accounts for about two and a half times the combined mass of all other planets in the solar system.

Composition and Atmosphere

Jupiter is a gas planet, like Saturn, Uranus, and Neptune, made up of 90% hydrogen and 10% helium. Although it is mostly gas, scientists believe it may have a solid core about the size of Earth. Deep inside Jupiter, hydrogen turns into liquid and then morphs into metal due to pressure. Jupiter has cloud belts that are red, brown, yellow, and white. These layers contain ammonia, sulfur, methane, and water vapor.

Large storms and hurricanes are constantly found on the surface of the planet. Jupiter has an interesting "Great Red Spot" that Giovanni Cassini, an Italian astronomer, spotted in 1665. The red spot is actually a 350-year-old storm, three times the size of Earth. Its reddish hue is still a topic of scientific investigation, with theories suggesting that it may result from chemical reactions in the atmosphere. Winds within the Great Red Spot can reach speeds of up to four hundred miles per hour.

Rings of Jupiter

Jupiter has four faint rings called the Halo ring, Main ring, Amalthea Gossamer ring, and the Thebe Gossamer ring. The gravitational influence of Jupiter helps maintain the structure and stability of these rings. The Voyager spacecraft first spotted three of the rings in 1976. In addition to the rings, Jupiter has the brightest auroras in the solar system. However, they can only be seen through ultraviolet light.

Jupiter rotates faster than any other planet. A day on Jupiter is only ten Earth hours. In fact, days on Jupiter are the shortest of all of the eight planets in the solar system. Jupiter is much further away from the Sun and takes a long time to make one revolution around it. One year on Jupiter takes 11.8 Earth years. Sunlight reaches Jupiter in about 45 minutes.

Moons of Jupiter

Unlike the rocky, terrestrial planets which have zero to two moons, Jupiter has over ninety moons. Galileo first observed three "stars" near Jupiter in 1610. On subsequent nights, he noticed they changed positions but always remained near Jupiter. Eventually, he identified a fourth object. He realized these "stars" were actually moons orbiting Jupiter. This discovery provided evidence against the geocentric model (Earth-centered universe). Galileo named the moons Europa, Ganymede, Callisto, and Io. These four moons are known as the Galilean Moons.

Europa is closely studied by scientists today because of the possibility that it may be able to support life. It has a thin atmosphere made up of mostly oxygen and has a very smooth surface. Ganymede is the ninth largest object and the largest moon in the solar system (larger than Mercury). It is the only moon in the solar system with its own magnetic field.

Callisto is Jupiter's second largest moon and is 2.6 times smaller than Earth. Callisto is the most heavily cratered body in the solar system. Io is the densest moon in the solar system and is covered in volcanoes and lava. The most famous volcano, Loki Patera, is one of the largest and has been continuously active. In 2023, 12 new moons were discovered around Jupiter.

Space Exploration

Radiation surrounds the planet and makes it difficult to send a spacecraft. However, there have been successful missions launched by NASA that have helped scientists learn about the planet. The Pioneer mission sent two spacecraft to gather information on the environment and radiation, to see if it is possible to send spacecraft in the future. In 1973, Pioneer 10 was the first spacecraft to get close-up images of the planet. The Voyager discovered a few moons and a ring around Jupiter. The Galileo spacecraft reached Jupiter in 1995 and was the first to enter the atmosphere, staying in orbit for eight years. The Juno orbiter was launched in 2016, gathering data on the planet, its atmosphere, and magnetic field.

PARAGRAPH CODE

After reading about **Jupiter**, head back to the reading and number ALL the paragraphs in the reading passage. Then, read each statement below and determine which paragraph **NUMBER** the statement can be found in. Paragraph numbers MAY be used more than one time or not at all. Follow the directions below to reveal the 4-digit code.

A Jupiter has an interesting "Great Red Spot" that Giovanni Cassini, an Italian astronomer, spotted in 1665. ☐

B Jupiter was named after the king of the Roman gods in ancient mythology. ☐

C Jupiter has four faint rings called the Halo ring, Main ring, Amalthea Gossamer ring, and the Thebe Gossamer ring. ☐

D This discovery provided evidence against the geocentric model (Earth-centered universe). ☐

E Ganymede is the ninth largest object and the largest moon in the solar system (larger than Mercury). ☐

F You could fit eleven Earths side by side within Jupiter's diameter. ☐

G Jupiter is a gas planet, like Saturn, Uranus, and Neptune, made up of 90% hydrogen and 10% helium. ☐

H Winds within the Great Red Spot can reach speeds of up to four hundred miles per hour. ☐

ELIMINATE ALL EVEN-NUMBERED paragraphs that you used as an answer. Record the remaining numbers (in the SAME order in which you recorded them above) in the boxes below.

☐ ☐ ☐ ☐

MYSTERY WORD

After reading about **Jupiter**, determine if each statement below is true or false. Color or shade the boxes of the **TRUE** statements. Next, unscramble the mystery word using the large letters of the **TRUE** statements.

Jupiter has over twenty faint rings. **H**	Jupiter is a gas planet, made up of 90% oxygen and 10% hydrogen. **O**	The Galileo spacecraft reached Jupiter in 1923. **N**	Sunlight reaches Jupiter in about 45 minutes. **T**
Jupiter rotates slower than any other planet. **B**	Io is the densest moon in the solar system. **A**	Jupiter is the largest planet in the solar system. **C**	Jupiter has the brightest auroras in the solar system. **F**
Jupiter shrinks two centimeters every year because it radiates intense heat. **P**	A day on Jupiter is only ten Earth hours. **R**	Days on Jupiter are the longest of all of the eight planets in the solar system. **G**	Callisto is Jupiter's second largest moon. **S**
Jupiter's mass is over 318 times bigger than Earth. **C**	The "Great Red Spot" was discovered by Giovanni Cassini. **E**	Galileo named the moons Europa, Ganymede, Callisto, and Io. **A**	Europa is the largest moon in the solar system. **I**

Unscramble the word using the large bold letters of only the **TRUE** statements.

SATURN

Saturn is the sixth planet from the Sun in the solar system. It is the second largest planet with the largest rings surrounding it. Saturn is bright enough to be seen from Earth without a telescope. Discovered in ancient times, Saturn was named after the Roman god of harvest and agriculture.

Around four million years ago, dust and gases swirled together to create Saturn. The Assyrians first recorded Saturn in the 8th century BC. Galileo spotted Saturn in the 1600s. He thought he was looking at three planets or a planet with handles. He even called the rings of Saturn "ears." In 1659, Dutch astronomer Christiaan Huygens discovered and identified the rings using a more advanced telescope.

Saturn's Composition and Atmosphere

Saturn is a gas giant meaning it has no solid surface. There is a solid core in the middle, possibly made of iron and nickel. Most of the planet is composed of gases, 94% hydrogen and 4% helium. It is the least dense planet in the solar system. Temperatures on Saturn are cold with an average temperature of -285 degrees Fahrenheit. It is very windy, and the weather can change drastically with sudden violent storms that can be seen from Earth. These storms create the appearance of stripes that are known as White Spots. Winds near the equator on Saturn can reach 1,118 miles per hour.

Rings and Moons

Saturn is considered the 'King of the Moons' with 146 confirmed moons. However, some astronomers think a few of those moons need more exploration or research. Christiaan Huygens discovered Titan, the largest moon on Saturn, in 1655. The moon Titan is second only to Jupiter's Ganymede moon. Titan is larger than the planet Mercury. Saturn also has a small moon called Mimas. The second largest moon of Saturn is Rhea, named after the mother of Olympian gods in Greek mythology. Dione is the largest inner moon and Saturn's fourth largest moon. It is only about 20% of the size of Titan. In 2023, the number of moons increased when sixty-two moons were discovered, bringing the total to over one-hundred-forty.

Although it is not the only planet with rings, Saturn's rings are the only ones that can be seen easily from Earth, even with a regular telescope. It is not known for sure how the rings formed. Scientists believe they may be pieces of comets, asteroids, and other objects that broke apart before reaching the planet, and Saturn's strong gravity pulled the pieces into its orbit. The rings span 250,000 miles.

Saturn is nicknamed the 'Ringed Planet' because it has very bright, visible rings. In this complex ring system, there are seven main rings. These rings are made up of chunks of rocks and ice, some very small and some larger than half a mile across. The gap between the rings is called the Cassini Division. Astronomers have labeled Saturn's main rings with letters of the alphabet. The main rings are called A, B, and C. Saturn's rings are thin with an average thickness of just thirty-three feet.

Beyond these main rings, there are additional fainter rings, including the D, E, F, and G rings. The rings are not solid but are instead composed of millions of individual particles orbiting Saturn in a disk-like formation. The E ring is primarily composed of ice particles ejected from the moon Enceladus. Enceladus has geysers that spew water ice and vapor from its subsurface ocean, feeding the E ring. The main rings also include smaller rings, so the exact number of rings is not known.

Size and Rotation

Saturn rotates very quickly with one day lasting only 10.7 Earth hours. Saturn has the second-shortest day of any planet in the solar system. However, a year on Saturn lasts over 29 Earth years. Due to its slow movement, the ancient Assyrians nicknamed the planet "Lubadsagush," which means "oldest of the old."

While Saturn is only a bit smaller than Jupiter, its mass is much smaller. Still, the mass of Saturn is ninety-five times Earth's mass. Saturn's surface area is eighty-three times larger than Earth's. Saturn appears to be yellow, caused by ammonia crystals in the upper atmosphere.

Space Exploration

There have been a few successful spacecraft missions to Saturn. The first to explore the planet was Pioneer 11, launched in 1973. Voyager 1 and Voyager 2, which launched in 1977, explored Saturn soon after Pioneer 11. This spacecraft took thousands of close-up photos of the planet and the moons which were the first photos that showed details of Saturn's rings.

In 2004, Cassini-Huygens became the first craft to orbit Saturn and study the planet's rings and moons. It also sent a probe to the moon, Titan, which gathered information and discovered the presence of water. The mission was a cooperative project of NASA, the European Space Agency (ESA), and the Italian Space Agency (ASI) aimed at studying Saturn.

MYSTERY MATCH

After reading about **Saturn**, draw a line from the left-hand column to make a match in the right-hand column. Your line should go through **ONE** letter. When you complete all the matches, use the letters with a line THROUGH them to unscramble a mystery word. You MUST start and end your line at the **arrow points**.

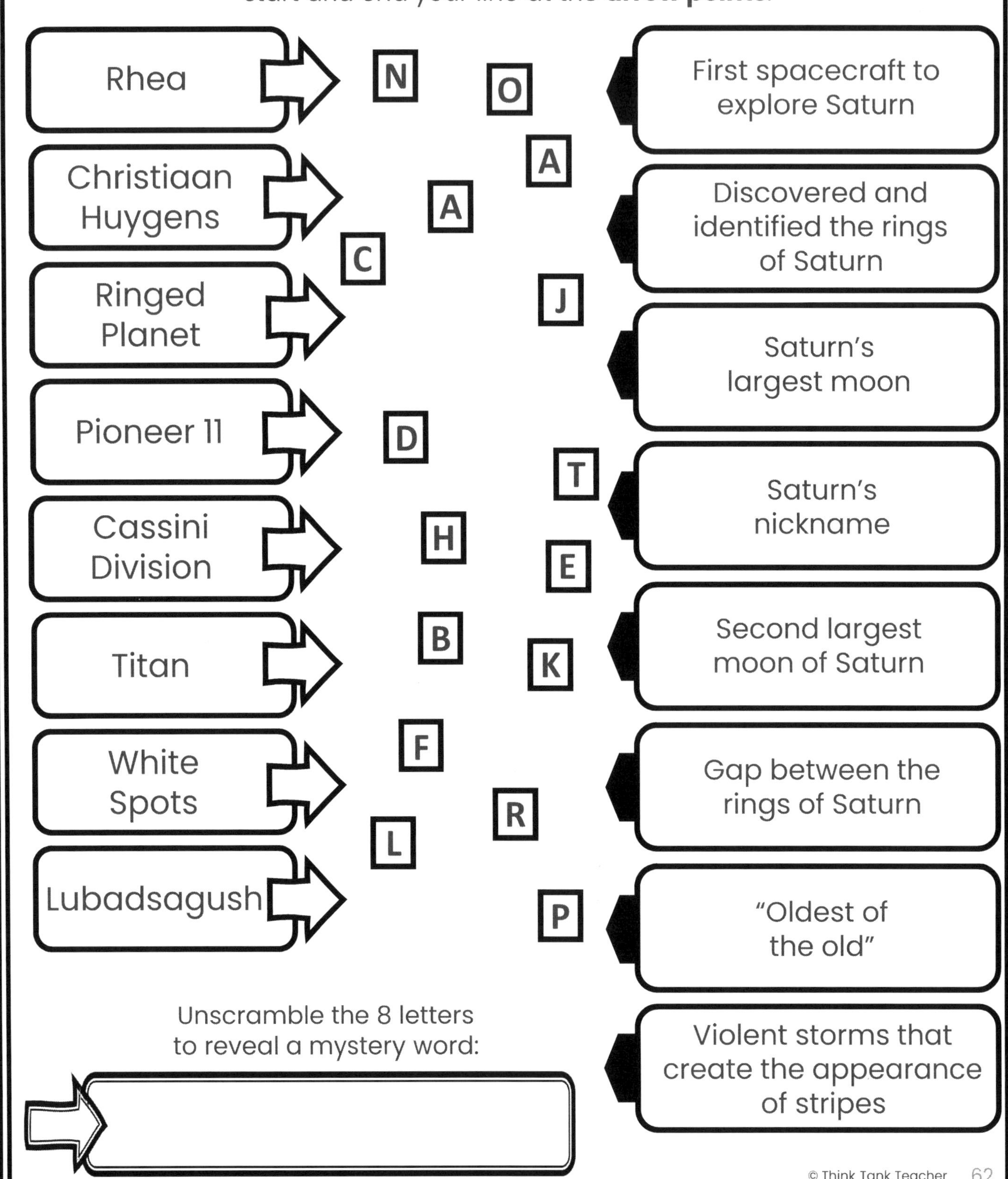

Unscramble the 8 letters to reveal a mystery word:

MULTIPLE CHOICE

After reading about **Saturn**, answer each multiple-choice question below. Then, count the number of times you used each letter as an answer (ABCD) to reveal a 4-digit code. Letters may be used more than once or not at all. If a letter option is not used, put a zero in that box.

1 How many Earth years is one year on Saturn?

A. 4 years
B. 16 years
C. 29 years
D. 43 years

2 The E ring is primarily composed of ice particles ejected from which moon?

A. Dione
B. Titan
C. Mimas
D. Enceladus

3 Which moon is the largest inner moon and Saturn's fourth largest moon?

A. Rhea
B. Dione
C. Titan
D. Mimas

4 Saturn was named after the Roman god of what?

A. Sun and sky
B. Harvest and agriculture
C. Love and beauty
D. Travel and thievery

5 Which ancient civilization was the first to record Saturn in the 8th century BC?

A. Aztec
B. Anglo-Saxons
C. Assyrians
D. None of the above

6 Due to ammonia crystals in the upper atmosphere, what color does Saturn appear to be?

A. Red
B. Green
C. Yellow
D. Blue

7 When were the Voyager 1 and Voyager 2 launched?

A. 1947
B. 1957
C. 1967
D. 1977

8 In Saturn's complex ring system, how many main rings are there?

A. Two
B. Seven
C. Fifteen
D. Over fifty

Count how many times you used each letter as a correct answer (ABCD) to determine the 4-digit code. Record your answer in the boxes below.

# of A's

# of B's

# of C's

# of D's

URANUS

Uranus is the seventh planet from the Sun and the third largest in the solar system. Due to the presence of methane, the clouds on the planet give Uranus a blue appearance. The discovery of Uranus doubled the size of the known solar system and challenged existing models of planetary motion. It was common to name a newly discovered element after a planet. In this case, Uranium was named after Uranus in 1789.

Discovery and Naming

John Flamstead first spotted Uranus in 1690; however, he thought it looked like a star. Astronomer William Herschel discovered Uranus by telescope on March 13, 1781. Yet, he thought it was only a comet. Finally, two years later, Herschel decided it was a planet. He tried to name the planet "Georgium Sidus" after King George III, the reigning king of England at the time. However, others did not like that name. Finally, astronomer Johan Bode suggested the name Uranus, after the Greek god Ouranos, the god of the sky and husband of the Earth. Uranus is the only planet named after a Greek god. Roman gods inspired the other planets' names.

Unique Tilt and Rotation

Uranus spins in the opposite direction of Earth and most other planets. It is the only planet in the solar system that is so tilted, it spins on its side. The extreme tilt of the planet at ninety-eight degrees has the Sun hitting parts of it for over forty years and keeping other parts of the planet in the dark, without heat. The cause of Uranus' unusual tilt is not fully understood, but it is widely believed to be the result of a colossal impact with an Earth-sized object early in the planet's history. This collision would have knocked the planet onto its side, creating the unique rotational dynamics observed today. With temperatures that can dip to -224 degrees Fahrenheit, it is impossible for life to exist. To reach Uranus, sunlight takes three hours.

Uranus spins faster than Earth and one rotation, or day, only takes 17 Earth hours. It takes the planet a long time to travel around the Sun. One year, or revolution, on Uranus takes just over 84 Earth years. The north pole seasons on Uranus have 21 years of nighttime in winter, 21 years of daytime in summer, and 42 years of day and night in the spring and fall. On the planet, powerful winds that can exceed 560 miles per hour have been recorded and large storms (bigger than the size of the United States) can take place.

Atmosphere and Composition

Uranus is a gas giant along with Jupiter, Saturn, and Neptune. However, because its composition is different from Jupiter and Saturn and the fact that it's made up of mostly icy material, it is considered an ice giant along with Neptune. Like all of the gas giants, Uranus does not have a solid surface and only a small rocky center. Out of the four gas giants, Uranus has the lowest mass.

Uranus has a thick atmosphere made of 82.5% hydrogen, 15.2% helium, and a small amount of methane. Methane (CH_4) is the third-most-abundant component of the atmosphere of Uranus.

Rings and Moons

There are thirteen known rings around Uranus. They are categorized as inner and outer rings. The inner rings are dark and narrow and much harder to see, while the outer rings are more brightly colored. The brightest ring of Uranus is the epsilon ring.

The rings were first discovered in 1977 by James Elliot, Edward Dunham, and Jessica Mink. Though, British astronomer William Herschel claimed to have made ring observations in 1789. Scientists believe the rings were created when moons crashed into each other and broke into many pieces that fell into its orbit.

Uranus has twenty-seven moons that have been discovered so far, with many more likely in the rings. Many moons are named after characters written by William Shakespeare and Alexander Pope. Some names include Miranda, Ariel, Umbriel, Puck, and Oberon. Even the biggest moon of Uranus is smaller than Australia.

One of the biggest moons, Umbriel, is very dark and mostly made from ice, with only a tiny portion from rock. The largest moon is called Titania. The moons are divided into three main categories: thirteen inner moons, nine irregular moons, and five major moons. The five major moons—Miranda, Ariel, Umbriel, Titania, and Oberon—are particularly notable for their varied and geologically active surfaces. Miranda, a major moon, has giant fault canyons that can reach twelve times deeper than the Grand Canyon. Of the major moons, it is the innermost and smallest. Miranda was discovered by Gerard Kuiper in 1948. Oberon is the outermost of the five major moons, while Ariel is the brightest Uranian moon.

TRUE OR FALSE

After reading about **Uranus**, read each statement below and determine if it is true or false. If the statement is true, color the coin that corresponds with that question. If the statement is false, cross out that coin value. When you are finished, add the TOTAL of **ALL TRUE** coin values to reveal a 4-digit code. One digit of the code has been provided for you. If the total is 625, a 6 would go in the first box, the 2 in the second box and so on.

A. Miranda was discovered by Johannes Kepler in 1643.

B. The inner rings are dark and narrow and much harder to see, while the outer rings are more brightly colored.

C. Uranus is the only planet named after a Greek god.

D. Uranus is the fourth planet from the Sun and the seventh largest in the solar system.

C
50

E. Uranus spins slower than Earth and one rotation, or day, takes 158 Earth hours.

F. Umbriel is the brightest Uranian moon.

G. The five major moons include Miranda, Ariel, Umbriel, Titania, and Oberon.

H. The discovery of Uranus doubled the size of the known solar system and challenged existing models of planetary motion.

After shading the coins based on your answer, add the value of ALL TRUE statements to get the final total. Record your answer in the boxes below.

7

DOUBLE PUZZLE

After reading about **Uranus**, determine the word that corresponds with the statements provided below. Spell the corresponding word in the boxes to the right. You may or may not use all squares provided for each answer. Any numerical answers must be spelled out. Next, use the numbers **under** indicated letters to reveal a secret word.

#	Statement	Answer	Number under letter
1	Last name of astronomer that suggested the name Uranus		8
2	The outermost of the five major moons of Uranus		2
3	Brightest ring of Uranus		6
4	The third-most-abundant component of the atmosphere of Uranus		9
5	Largest moon of Uranus		5
6	Number of hours it takes sunlight to reach Uranus		
7	Due to the presence of methane, Uranus appears this color		3
8	Number of moons classified as irregular moons of Uranus		7
9	Last name of British astronomer who claimed to make ring observations in 1789		1
10	Brightest Uranian moon		4

SECRET WORD

1	2	3	4	5	6	7	8	9

NEPTUNE

Neptune is the eighth and most distant planet in the solar system. It is the fourth largest planet and the furthest from the Sun. In fact, Neptune is thirty times further from the Sun than Earth is. Neptune is so far from Earth that it is impossible to see without the use of a telescope. Neptune was named after the Roman god of the sea due to its blue appearance.

Discovery and Naming

There is some controversy over who discovered Neptune. Galileo is thought to be the first person to view Neptune through a telescope in 1613, but he thought it was a star. It was not discovered as a planet until 1846 by Urbain Le Verrier, John Couch Adams, and Johann Galle. Astronomers, at first, determined that Uranus' orbit around the Sun did not match their predictions. So, they assumed that another planet was pulling on Uranus with gravity. Now astronomers understand that Neptune has the second-largest gravity of the planets.

The discovery of Neptune was not just a triumph for Le Verrier and Galle; it represented a turning point in astronomy. It showcased the power of mathematics and theoretical models in predicting celestial phenomena.

Classification, Composition, and Rotation

Neptune is the smallest of the four gas giants which also include Jupiter, Saturn, and Uranus. Because its composition is different from Jupiter and Saturn, Neptune is considered an ice giant along with Uranus. Like Uranus, the blue color comes from the methane in the atmosphere. Methane absorbs red light from the Sun. However, the blue light gets reflected back into space, so Neptune appears blue. Similar to all the gas giants, Neptune does not have a solid surface. Many consider Neptune to be Uranus' twin due to its comparable size and composition.

Neptune spins, or rotates, quickly and one day takes 16 Earth hours. Being the farthest planet from the Sun, it takes the longest to make a complete revolution around it. One year on Neptune is 165 Earth years.

Neptune has four layers; the core, mantle, lower atmosphere, and upper atmosphere. The upper atmosphere of clouds contain about 80% hydrogen and 19% helium. Likewise, the lower atmosphere is composed of hydrogen, helium, and methane. The mantle is made up of water, ammonia, and methane ice. Lastly, the core is composed of rock and ice. “Icy” materials such as water, ammonia, and methane make up roughly 80% of Neptune’s mass. Despite its great distance from the Sun, Neptune radiates more heat than it receives. Extreme pressure and temperatures mean that life cannot survive on Neptune.

Moons and Rings of Neptune

Neptune has fourteen official moons, with many more likely existing. The moons are all named after gods of water. Triton is the largest of the moons and orbits Neptune in the opposite direction. It is the only large moon in the solar system with a retrograde orbit. Triton is the seventh-largest moon of all the planets. Research confirms that Triton has active geysers that spew up icy nitrogen frost onto the surface. After the discovery of Neptune, Triton was discovered just seventeen days later.

Proteus is the second largest of Neptune's moons. This moon was discovered by Voyager 2 during its 1989 flyby. Neptune's inner moons - Naiad, Thalassa, Despina, Galatea, and Larissa - form a closely-knit group that orbits the planet within its ring system. The moon closest to the planet is Naiad. The fourteenth moon of Neptune, named Hippocamp, was discovered in 2013 while analyzing images collected by the Hubble Space Telescope.

The rings of Neptune are smaller and less visible than the rings on other planets and were not discovered until the 1980s. Neptune has six rings and four ring arcs. Some ring names include Galle, Le Verrier, Lassell, Arago, and Adams. Astronomers believe Neptune's rings were most likely formed when one of Neptune's moons was destroyed.

Weather and Winds

Neptune is a cold, dark and windy planet with average surface temperatures of -353 degrees Fahrenheit. Temperatures are extremely cold, and the weather is very active and more intense than on any other planet. The strongest and most powerful winds in the solar system are on planet Neptune. Winds can reach speeds greater than 1,200 mph, which is almost the same speed as the Hornet fighter jet in the U.S. Navy. Earth's most powerful winds have only hit 250 mph.

Dark spots can be seen in the atmosphere of Neptune which are exceptionally large storms. In 1989, Voyager 2 spotted one of the most significant storms ever recorded. Neptune's storm, called the "Great Dark Spot," was the size of Earth and lasted five years. In 1994, the Hubble Telescope collected data showing that the Great Dark Spot had disappeared, but another dark spot was later discovered.

PARAGRAPH CODE

After reading about **Neptune**, head back to the reading and number ALL the paragraphs in the reading passage. Then, read each statement below and determine which paragraph **NUMBER** the statement can be found in. Paragraph numbers MAY be used more than one time or not at all. Follow the directions below to reveal the 4-digit code.

A Neptune is the smallest of the four gas giants which also include Jupiter, Saturn, and Uranus.

B Triton is the largest of the moons and orbits Neptune in the opposite direction.

C Winds can reach speeds greater than 1,200 mph, which is almost the same speed as the Hornet fighter jet in the U.S. Navy.

D Extreme pressure and temperatures mean that life cannot survive on Neptune.

E After the discovery of Neptune, Triton was discovered just seventeen days later.

F The upper atmosphere of clouds contain about 80% hydrogen and 19% helium.

G Neptune spins, or rotates, quickly and one day takes 16 Earth hours.

H Neptune was named after the Roman god of the sea due to its blue appearance.

ELIMINATE ALL EVEN-NUMBERED paragraphs that you <u>used</u> as an answer. Record the remaining numbers (in the SAME order in which you recorded them above) in the boxes below.

MYSTERY WORD

After reading about **Neptune**, determine if each statement below is true or false. Color or shade the boxes of the **TRUE** statements. Next, unscramble the mystery word using the large letters of the **TRUE** statements.

Neptune is considered an ice giant along with Uranus. **E**	Neptune spins, or rotates, slowly and one day takes 681 Earth hours. **B**	The upper atmosphere of Neptune's clouds contain about 80% oxygen. **R**	Neptune is a cold, dark and windy planet. **S**
Ganymede is the second largest of Neptune's moons. **G**	Neptune's storm, called the "Great Dark Spot," lasted five years. **T**	Methane absorbs red light from the Sun. **E**	Hippocamp is the largest of Neptune's moons. **D**
Copernicus is thought to be the first person to view Neptune through a telescope in 1543. **A**	Neptune is thirty times further from the Sun than Earth is. **E**	The mantle is made up of water, ammonia, and methane Ice. **O**	Neptune has fourteen official moons, with many more likely existing. **L**
Neptune was named after the Greek god of the sky because it appears blue. **U**	Neptune has the second-largest gravity of the planets. **C**	Neptune has eight rings and twelve ring arcs. **H**	There is some controversy over who discovered Neptune. **P**

Unscramble the word using the large bold letters of only the **TRUE** statements.

PLUTO & DWARF PLANETS

Pluto, once the ninth planet in the solar system, is a dwarf planet that orbits the Sun. It is located in the Kuiper Belt, a region beyond the eighth planet of Neptune, which contains icy objects. Seeing Pluto from Earth with the naked eye is impossible, and even with a telescope it can be difficult.

When Pluto was first discovered in 1930 by Clyde Tombaugh, it was named the ninth planet. For seventy-five years, it held that title. However, in 2006, the International Astronomical Union, an organization of scientists from all over the world, changed its classification to a dwarf planet.

Pluto's location was first predicted by Percival Lowell in 1915, who hypothesized the existence of a "Planet X" beyond Neptune. Tombaugh discovered Pluto using a telescope in Flagstaff, Arizona. An eleven-year-old girl from England, named Venetia Burney, named Pluto after the Roman god of the underworld. In Greek mythology, Pluto is named after Hades. Pluto was considered the son of Saturn. Saturn ruled the world alongside his brothers: Jupiter controlled the sky, Neptune controlled the sea, and Pluto controlled the underworld. The name was fitting, given Pluto's dark and distant nature.

Reclassification

In 2006, another object was found in the solar system that was about the same size as Pluto. This object became known as Eris. This object was much smaller than the eight planets but still had similar characteristics. This led the International Astronomical Union to give an official definition for what objects are considered full sized planets, and Pluto no longer qualified. The planet must be spherical, orbit the Sun, and have a gravitational force that clears other objects from its orbit. Pluto checks off the first two criteria but not the third because it has other objects in its orbit.

Composition and Features

Pluto is smaller than Earth's moon. Pluto mainly consists of 70 percent rock and 30 percent water ice (nitrogen ice) with mountains, valleys, and craters covering its surface. There are a few areas of Pluto that are distinct, such as Tombaugh Regio, a very large and bright spot. The Brass Knuckles are dark areas of Pluto, located near the center of the planet.

There are five known moons that orbit Pluto: Charon, Styx, Nix, Hydra and Kerberos. Styx is the smallest moon of Pluto and was not discovered until 2012. Charon, discovered in 1978, is the largest of the moons and the largest known moon of any dwarf planet. Pluto's furthest moon is Hydra, and its closest moon is Charon. Some experts call Pluto and Charon a double planet instead of a planet and its moon. Both Nix and Hydra were discovered by the Hubble Space Telescope in 2005. These moons are irregularly shaped and likely formed from debris resulting from a collision between Pluto and another Kuiper Belt object.

Pluto's thin atmosphere (like a comet) is 98 percent nitrogen with trace amounts of methane and carbon monoxide. Astronomers believe the reddish color of Pluto is from hydrocarbon molecules.

Pluto spins slowly, tilted almost on its side. One rotation, or day, takes 153 Earth hours. It takes Pluto 248 Earth years to complete one revolution around the Sun. When it is closer to the Sun, the atmosphere becomes very thin.

Pluto does not receive much sunlight, so it is very cold, with an average temperature of -387 degrees Fahrenheit. It takes sunlight over five hours to reach Pluto. Life would be impossible on Pluto.

Recognized Dwarf Planets

Although Pluto was officially named a dwarf planet by the IAU, some scientists believe the new definition of a planet is inaccurate and Pluto should still be considered one of the planets. Despite its reclassification, Pluto remains a significant object of study. Pluto provides clues about the processes that shaped the early solar system. Its position in the Kuiper Belt and its composition offers insights into the formation and migration of planets. Evidence of geological activity on Pluto, such as tectonics and potential cryovolcanism, challenges our understanding of geological processes on icy bodies.

Pluto and other similar objects are now known as dwarf planets. There are currently five dwarf planets: Pluto, Eris, Ceres, Makemake, and Haumea. Of the five, Pluto is the largest. It is the second closest to the Sun, with Ceres being the closest.

Ceres is the only dwarf planet located in the asteroid belt between Mars and Jupiter. Discovered in 1801, Ceres is the smallest recognized dwarf planet. Eris is slightly smaller than Pluto and lies in the Kuiper Belt. It was discovered in 2005, sparking the debate that led to Pluto’s reclassification. Eris has one known moon, Dysnomia. Haumea is recognized for its oval shape caused by its rapid rotation. It is located in the Kuiper Belt, surrounded by a ring system with two moons, Hiʻiaka and Namaka. Makemake was discovered in 2005. It has a methane-covered surface, and one recently discovered moon.

MYSTERY MATCH

After reading about **Pluto & Dwarf Planets**, draw a line from the left-hand column to make a match in the right-hand column. Your line should go through **ONE** letter. When you complete all the matches, use the letters with a line THROUGH them to unscramble a mystery word. You MUST start and end your line at the **arrow points**.

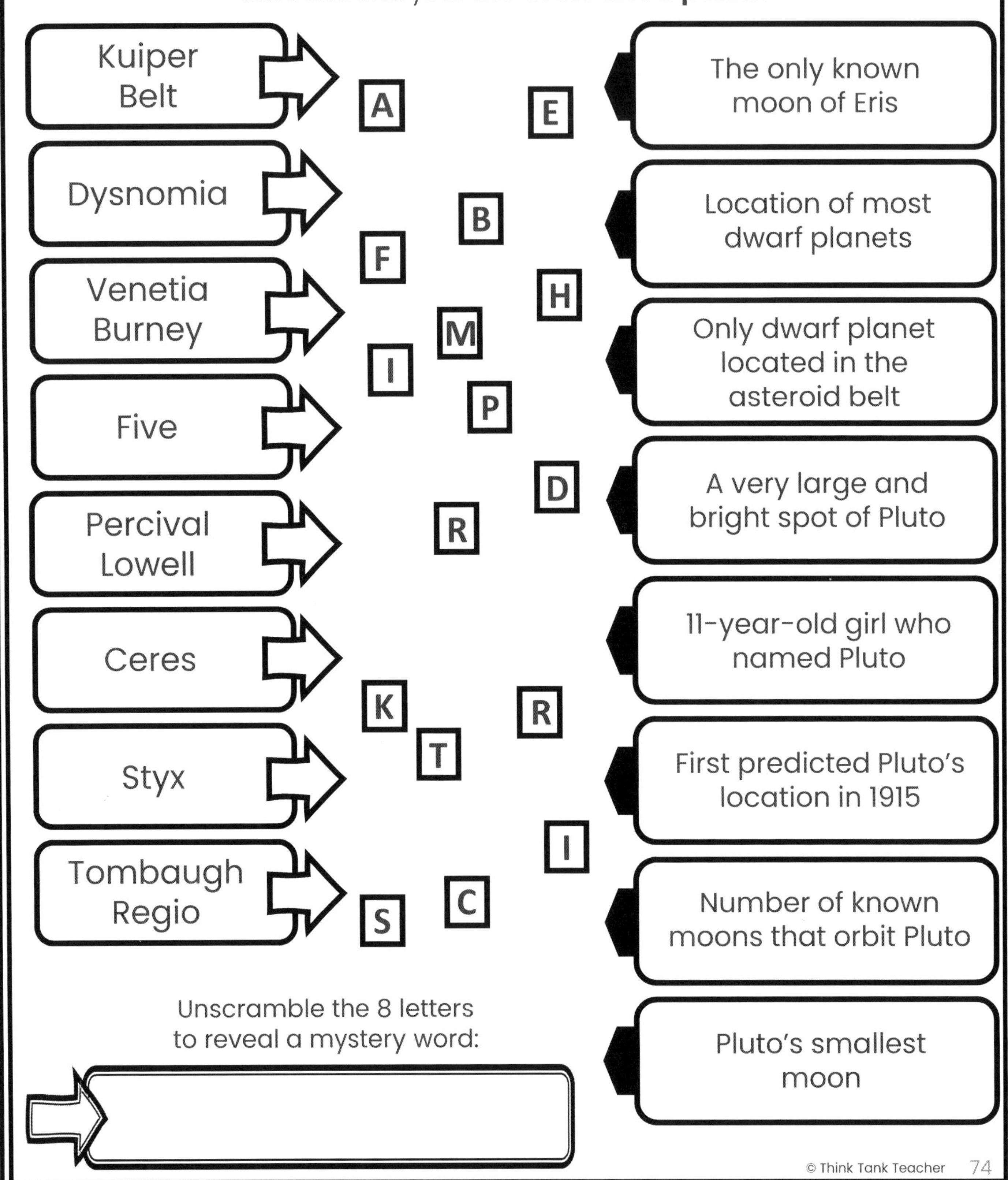

Unscramble the 8 letters to reveal a mystery word:

MULTIPLE CHOICE

After reading about **Pluto & Dwarf Planets**, answer each multiple-choice question below. Then, count the number of times you used each letter as an answer (ABCD) to reveal a 4-digit code. Letters may be used more than once or not at all. If a letter option is not used, put a zero in that box.

1 What covers the surface of Pluto?

A. Mountains
B. Craters
C. Valleys
D. All of the above

2 What is Pluto's largest moon?

A. Charon
B. Styx
C. Eris
D. Hydra

3 How many Earth years does it take Pluto to complete one revolution around the Sun?

A. 153 years
B. 248 years
C. 317 years
D. Over 500 years

4 When did the International Astronomical Union change Pluto's classification to a dwarf planet?

A. 2000
B. 2002
C. 2004
D. 2006

5 When was Pluto first discovered by Clyde Tombaugh?

A. 1930
B. 1940
C. 1950
D. 1960

6 Pluto was named after the Roman god of what?

A. Sea
B. Underworld
C. Sky
D. Mountains

7 Which of the following is not a dwarf planet?

A. Ceres
B. Makemake
C. Nix
D. Haumea

8 Due to hydrocarbon molecules, what color does Pluto appear?

A. Green
B. Red
C. Yellow
D. None of the above

Count how many times you used each letter as a correct answer (ABCD) to determine the 4-digit code. Record your answer in the boxes below.

# of A's	# of B's	# of C's	# of D's

SMALL CELESTIAL BODIES

ASTEROIDS

Asteroids are small, rocky remnants (leftovers) from the early formation of our solar system about 4.6 billion years ago. Located primarily in the Asteroid Belt, asteroids vary widely in size, composition, and location.

The term "asteroid" is derived from the Greek word "asteroeides," which means "star-shaped." This name reflects the appearance of asteroids when viewed through early telescopes. To astronomers of the 19th century, these celestial objects appeared as tiny points of light, much like stars, because of their small size and distance from Earth.

The word "asteroid" was first introduced in 1802 by British astronomer William Herschel, who used it to describe the newly discovered minor planets like Ceres and Pallas. Herschel distinguished these objects from planets due to their lack of visible disks, unlike the larger, rounder planets observed in the night sky. Over time, the term became standard for describing the small rocky bodies within our solar system.

Formation of Asteroids

Asteroids, sometimes referred to as minor planets, are often described as leftover building blocks from the solar system's formation. When the Sun and planets formed from a rotating disk of gas and dust, not all material merged into planets. In regions where gravitational forces were too disruptive, particularly near Jupiter, material failed to form a planet and instead became smaller bodies known as asteroids.

The Asteroid Belt, located between the orbits of Mars and Jupiter, contains the majority of known asteroids. It spans a vast region approximately 140 million miles wide and is thought to contain millions of asteroids. The objects within the Asteroid Belt are separated by significant distances, making collisions rare. This region's formation is linked to the gravitational influence of Jupiter, whose powerful pull prevented the rocky material from forming into a planet.

Largest Asteroids in the Solar System

Ceres is the largest object in the Asteroid Belt and the only one classified as a dwarf planet. It spans 587 miles in diameter and contains about 40% of the Asteroid Belt's total mass. Ceres was first discovered by Italian astronomer Giuseppe Piazzi in 1801. Ceres was named after the Roman goddess of the harvest.

Vesta is the second-largest asteroid, with a diameter of about 329 miles.

Vesta is unique due to its differentiated structure, with a crust, mantle, and core, resembling a small planet. Vesta is the brightest asteroid when viewed from Earth. Pallas is the third-largest asteroid, approximately 318 miles wide. Its orbit is unusually inclined compared to most asteroids. With a diameter of about 270 miles, Hygiea is the fourth-largest asteroid and notable for its nearly spherical shape.

Composition of Asteroids

Asteroids are classified into three main types based on their composition. C-Type (Carbonaceous) are the most common, making up about 75% of known asteroids. S-Type (Silicaceous), also known as Stony asteroids, are composed mostly of silicate rocks and metals. These asteroids are brighter and account for about 17% of known asteroids. M-Type (Metallic) are rare asteroids made mostly of metals like nickel and iron, possibly remnants of larger bodies whose rocky mantles were stripped away by collisions.

Location and Distribution

While the Asteroid Belt is the primary region for asteroids, they can also be found in other locations. Near-Earth Asteroids (NEAs) have orbits that bring them close to Earth's orbit. They are of particular interest due to their potential for impacting Earth. Trojan Asteroids share an orbit with a larger planet, such as Jupiter, but remain stable due to gravitational forces. Trans-Neptunian Objects are distant asteroids, like those in the Kuiper Belt, which reside far beyond Neptune's orbit and share characteristics with comets.

Monitoring programs such as NASA's Planetary Defense Coordination Office track potentially hazardous asteroids to mitigate these risks. Global networks like the International Asteroid Warning Network (IAWN) also closely monitor asteroids. Advanced telescopes and missions like OSIRIS-REx and DART (Double Asteroid Redirection Test) help scientists understand their behavior and develop strategies for impact mitigation.

Exploration of Asteroids

Humanity's understanding of asteroids has grown significantly thanks to space missions. The Dawn Mission spacecraft explored Vesta and Ceres, providing detailed images and data about their surfaces and compositions. The Hayabusa2 and OSIRIS-REx missions have collected and returned samples from asteroids Ryugu and Bennu. Bennu is a small Near-Earth Asteroid (NEA) discovered in 1999 that follows an orbital path bringing it close to Earth approximately once every six years.

TRUE OR FALSE

After reading about **Asteroids**, read each statement below and determine if it is true or false. If the statement is true, color the coin that corresponds with that question. If the statement is false, cross out that coin value. When you are finished, add the TOTAL of **ALL TRUE** coin values to reveal a 4-digit code. One digit of the code has been provided for you. If the total is 625, a 6 would go in the first box, the 2 in the second box and so on.

A. The Asteroid Belt is located between the orbits of Mercury and Venus.

B. Ceres was first discovered by Italian astronomer Giuseppe Piazzi in 1801.

C. S-Type (Silicaceous) are the most common, making up about 75% of known asteroids.

B 25

D. Pallas is the third-largest asteroid, approximately 318 miles wide.

E. The objects within the Asteroid Belt are separated by significant distances, making collisions rare.

C 50

F. Located primarily in the Asteroid Belt, asteroids vary widely in size, composition, and location.

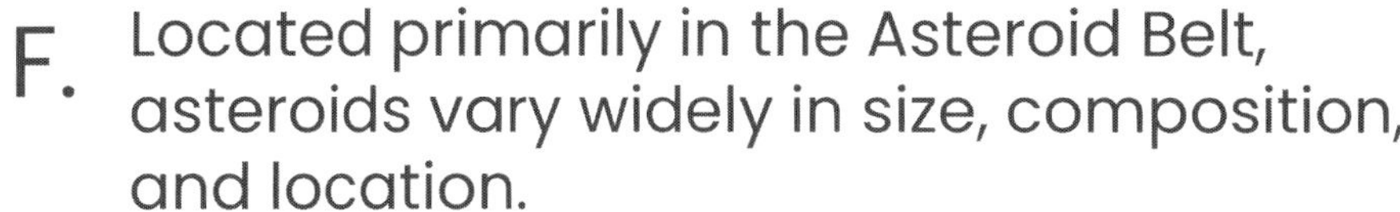

D 100

G. The word "asteroid" was first introduced in 1802 by Italian astronomer Galileo Galilei.

H. M-Type (Metallic) are rare asteroids made mostly of metals like nickel and iron.

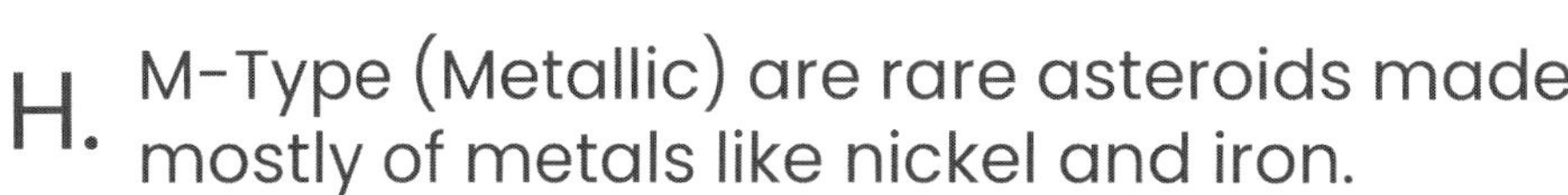

After shading the coins based on your answer, add the value of ALL TRUE statements to get the final total. Record your answer in the boxes below.

DOUBLE PUZZLE

After reading about **Asteroids**, determine the word that corresponds with the statements provided below. Spell the corresponding word in the boxes to the right. You may or may not use all squares provided for each answer. Any numerical answers must be spelled out. Next, use the numbers **under** indicated letters to reveal a secret word.

1	The brightest asteroid when viewed from Earth	☐☐☐☐☐☐☐☐☐ (4th box: 5)
2	The Asteroid Belt is located between the orbits of Mars and ___	☐☐☐☐☐☐☐☐☐
3	A small Near-Earth Asteroid (NEA) discovered in 1999	☐☐☐☐☐☐☐☐☐ (1st box: 3)
4	International Asteroid ___ Network (IAWN)	☐☐☐☐☐☐☐☐☐ (2nd box: 6)
5	Asteroids are classified into ___ (#) main types	☐☐☐☐☐☐☐☐☐
6	___ Asteroids share an orbit with a larger planet but remain stable	☐☐☐☐☐☐☐☐☐ (2nd box: 2)
7	S-Type (Silicaceous) are also known as ___ asteroids	☐☐☐☐☐☐☐☐☐ (3rd box: 1)
8	Fourth-largest asteroid	☐☐☐☐☐☐☐☐☐
9	Last name of Italian astronomer that first discovered Ceres	☐☐☐☐☐☐☐☐☐ (2nd box: 4)
10	Asteroids are sometimes referred to as minor ___	☐☐☐☐☐☐☐☐☐ (2nd box: 7)

SECRET WORD

☐	☐	☐	☐	☐	☐	☐
1	2	3	4	5	6	7

COMETS

Comets are among the most fascinating celestial objects in the solar system. Often described as "dirty snowballs," they provide insights into the early solar system's formation. Composed of rock, dust, ice, and frozen gases, comets follow elongated orbits that bring them close to the Sun, where their tails and comas become visible.

In ancient times, the appearance of a comet in the sky was often associated with doom, foretelling natural disasters, the downfall of rulers, impending wars, or significant changes. These beliefs were fueled by the lack of scientific understanding of comets' nature and origins. Seen as fiery streaks cutting across the sky, comets were thought to reside in the realm of the gods, carrying divine messages or warnings to Earth.

Structure of a Comet

Comets consist of three main parts: the nucleus, coma, and tails. As a comet approaches the Sun, the increasing heat causes dramatic changes to its structure. The nucleus is the solid core of a comet made up of a mixture of dust, rock, water ice, and frozen gases such as carbon dioxide, carbon monoxide, methane, and ammonia. Images from spacecraft have revealed irregular shapes, craters, and rough terrain on comet nuclei.

When a comet approaches the Sun, the heat causes sublimation - frozen gases turning directly into vapor. This process releases gas and dust, forming a bright, glowing cloud around the nucleus called the coma. The coma reflects sunlight and may glow faintly due to chemical reactions caused by solar radiation.

A comet's tail is formed from plasma, a state of matter where gases are so intensely heated that their atoms lose electrons, becoming electrically charged. As the comet gets closer to the Sun, solar wind and radiation pressure cause some of the released materials to stream away from the nucleus, forming a tail.

Comets develop two types of tails. An Ion Tail is composed of charged particles (ions) and the tail is pushed away from the Sun by the solar wind. It often appears blue due to ionized gases. Because Ion Tails are influenced by the solar wind and magnetic fields rather than gravity, they always point directly away from the Sun, regardless of the comet's direction of travel. A Dust Tail is made of small solid particles such as silicates and carbon-based materials, that are released from the comet's nucleus. The particles reflect sunlight leading to a yellow or white appearance. Dust Tails curve due to gravity and solar pressure.

Notable Comets

Halley's Comet is perhaps the most famous comet, visible from Earth approximately every 76 years. It was last seen in 1986 and will return in 2061. This comet was named after astronomer Edmond Halley, who correctly predicted its periodic return in 1705. During its 1986 approach, spacecraft like Giotto and Vega captured close-up images of its nucleus.

The Hale-Bopp Comet was one of the brightest comets of the 20th century and was visible to the naked eye for eighteen months in 1997. The Comet Shoemaker-Levy 9 made headlines in 1994 when it collided with Jupiter. The collision revealed details about Jupiter's atmosphere.

The European Space Agency's Rosetta spacecraft orbited Comet 67P/Churyumov-Gerasimenko in 2014 and deployed the Philae lander to its surface. The mission revealed the comet's complex surface, diverse chemical composition, and mechanisms of outgassing, revolutionizing our understanding of comets.

Formation and Location of Comets

Comets are typically categorized into two primary groups based on the length and nature of their orbits: short-period comets and long-period comets. Short-period comets are defined by their relatively quick orbits around the Sun, taking less than two hundred years to complete a full cycle.

Long-period comets, on the other hand, take over two hundred years, and often thousands or even millions of years to complete an orbit. These comets have unusual and elongated orbits, often traveling far beyond the main regions of the solar system.

Comets are believed to originate from two main regions in the solar system: the Kuiper Belt and Oort Cloud. The Kuiper Belt is a region beyond Neptune's orbit, containing icy bodies and dwarf planets. Comets from the Kuiper Belt, such as Halley's Comet, have orbital periods of less than two hundred years.

The Oort Cloud is located between 2,000 and 100,000 astronomical units (AU) from the Sun, making it the farthest-reaching structure in the solar system. Gravitational disturbances, such as interactions with nearby stars or the Milky Way's galactic tide, can occasionally dislodge an object from the Oort Cloud and send it on a trajectory toward the inner solar system, transforming it into a long-period comet. Comets from the Oort Cloud, such as Comet Hale-Bopp, have orbits lasting thousands to millions of years.

PARAGRAPH CODE

After reading about **Comets**, head back to the reading and number ALL the paragraphs in the reading passage. Then, read each statement below and determine which paragraph **NUMBER** the statement can be found in. Paragraph numbers MAY be used more than one time or not at all. Follow the directions below to reveal the 4-digit code.

A An Ion Tail is composed of charged particles (ions) and the tail is pushed away from the Sun by the solar wind.

B Often described as "dirty snowballs," they provide insights into the early solar system's formation.

C The Kuiper Belt is a region beyond Neptune's orbit, containing icy bodies and dwarf planets.

D Halley's Comet is perhaps the most famous comet, visible from Earth approximately every 76 years.

E The Comet Shoemaker-Levy 9 made headlines in 1994 when it collided with Jupiter.

F This comet was named after astronomer Edmond Halley, who correctly predicted its periodic return in 1705.

G Comets consist of three main parts: the nucleus, coma, and tails.

H This process releases gas and dust, forming a bright, glowing cloud around the nucleus called the coma.

ELIMINATE ALL EVEN-NUMBERED paragraphs that you used as an answer. Record the remaining numbers (in the SAME order in which you recorded them above) in the boxes below.

MYSTERY WORD

After reading about **Comets**, determine if each statement below is true or false. Color or shade the boxes of the **TRUE** statements. Next, unscramble the mystery word using the large letters of the **TRUE** statements.

The bright, glowing cloud around the nucleus is called the oort. **T**	Comets develop four types of tails. **P**	The Hale-Bopp Comet was one of the brightest comets of the 20th century. **G**	Sublimation is when frozen gases turn directly into vapor. **L**
The nucleus is the liquid core of a comet. **B**	When a comet approaches the Sun, the heat causes rain. **F**	Short period comets always take less than fifty years to complete a full cycle. **A**	Halley's Comet is visible from Earth approximately every 76 years. **N**
Dust Tails often appear blue due to ionized gases. **E**	Comets are composed of rock, dust, ice, and frozen gases. **O**	Halley's Comet collided with Jupiter in 1994. **R**	Comets are believed to originate from two main regions in the solar system. **W**
Comets consist of six main parts. **C**	The Kuiper Belt is a region beyond Neptune's orbit. **G**	A comet's tail is formed from oxygen. **D**	Short-period comets are defined by their quick orbits around the Sun. **I**

Unscramble the word using the large bold letters of <u>only</u> the **TRUE** statements.

METEORS

Meteoroids, meteors, and meteorites are related phenomena originating from the debris of celestial bodies like asteroids and comets. A meteoroid is a small particle traveling through space, ranging in size from dust grains to small boulders. When a meteoroid enters Earth's atmosphere, it heats up due to friction with air molecules, producing a bright streak of light known as a meteor. If any part of the meteoroid survives this fiery descent and reaches Earth's surface, it becomes a meteorite. These objects not only create spectacular light displays, such as meteor showers, but they also provide critical insights into the formation of the solar system. The word "meteor" originates from the Greek word meteōros, which means "high in the air."

Differences

Meteoroids are small, rocky or metallic objects in space, smaller than asteroids but larger than interplanetary dust. They originate from asteroids, comets, or collisions between celestial bodies. Ranging in size, meteoroids travel through space at astonishing speeds, at thousands of miles per hour.

When a meteoroid enters Earth's atmosphere, it encounters intense friction with air molecules. This interaction generates heat, causing the meteoroid to glow brightly and produce a streak of light across the sky. This phenomenon is called a meteor, or a "shooting star." Most meteors burn up entirely before reaching the surface of Earth.

If a meteoroid impacts Earth's surface, it is called a meteorite. Meteorites are classified into three main types: stony, metallic, and stony-iron, each with unique properties and origins. Stony meteorites are most commonly observed falling to Earth. Stony-iron meteorites are very rare.

Elements and Colors

Meteor colors are determined by two main factors: composition and speed. Different elements in the meteoroid emit distinct colors when they are vaporized due to intense heat as they enter Earth's atmosphere. These elements release specific wavelengths of light. The high velocity of the meteoroid and the interaction with Earth's atmosphere contribute to ionization, amplifying the colors produced. Sodium, a common element in meteors, produces a bright yellow or orange glow. Magnesium is responsible for a bluish-green hue when ionized. Silicon contributes to reddish or orange tints in the trail.

Meteor Showers

Meteor showers occur when Earth passes through debris left by comets or asteroids. As the planet moves through this trail of particles, numerous meteoroids enter the atmosphere, creating a flurry of meteors visible from the ground.

Meteor showers are named after the constellation or star nearest to their radiant, the point in the sky where the meteors appear to originate. For example, the Perseids are named after the constellation Perseus. Perseids are active in August, associated with the Swift-Tuttle Comet, and known for its bright, fast meteors. Leonids occur in November, tied to Tempel-Tuttle Comet, and known for spectacular meteor storms approximately every thirty-three years. Geminids are active in December, unusual for originating from an asteroid (3200 Phaethon) rather than a comet, producing multicolored meteors.

Impacts and Historical Significance

The Chicxulub Crater was a massive meteorite impact sixty-six million years ago near the Yucatán Peninsula and is widely believed to have caused the extinction of the dinosaurs. During the Tunguska Event in 1908, a large meteoroid exploded over Siberia, flattening eight-hundred square miles of forest. Though no impact crater was found, it remains one of the most powerful meteoroid explosions in recorded history.

Discovery

The Hoba West Meteorite was discovered in Namibia in 1920. It is the largest known meteorite, weighing about sixty tons. Composed primarily of iron (84%) and nickel (16%), it is classified as an ataxite. Ataxites are a rare class of iron meteorites. Remarkably, it did not create an impact crater, likely due to its flat shape and low entry velocity, which allowed it to land without a catastrophic impact.

El Chaco in Argentina is the second largest meteorite on Earth weighing thirty-seven tons. A fragment of the Cape York Meteorite, known as Ahnighito, was discovered in Greenland, weighing thirty-four tons. The Willamette Meteorite was found in Oregon and is the largest iron meteorite found in the United States and the sixth largest in the world. The Murchison Meteorite fell in Australia in 1969 and is known for containing organic compounds. The Allende Meteorite fell in Mexico in 1969 and was found to contain amino acids, hinting at the building blocks of life arriving on Earth via space debris.

MYSTERY MATCH

After reading about **Meteors**, draw a line from the left-hand column to make a match in the right-hand column. Your line should go through **ONE** letter. When you complete all the matches, use the letters with a line THROUGH them to unscramble a mystery word. You MUST start and end your line at the **arrow points**.

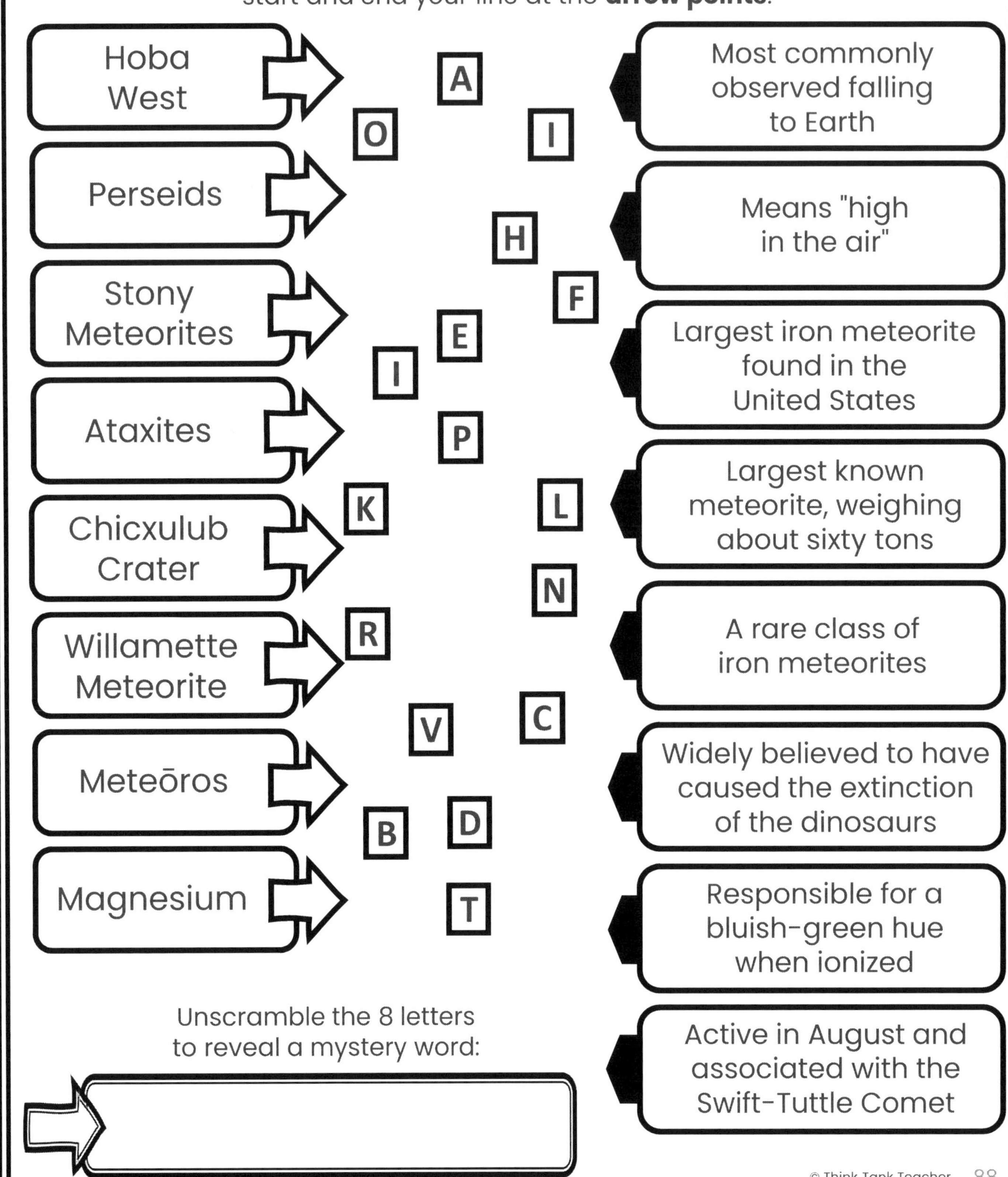

Unscramble the 8 letters to reveal a mystery word:

MULTIPLE CHOICE

After reading about **Meteors**, answer each multiple-choice question below. Then, count the number of times you used each letter as an answer (ABCD) to reveal a 4-digit code. Letters may be used more than once or not at all. If a letter option is not used, put a zero in that box.

1 Where was the Hoba West Meteorite discovered?

A. Siberia
B. Oregon
C. Mexico
D. Namibia

2 What constellation were the Perseids named after?

A. Perseus
B. Pegasus
C. Poseidon
D. Prometheus

3 Meteorites are classified into how many main types?

A. Three
B. Six
C. Nine
D. Twelve

4 What is the second largest meteorite on Earth weighing thirty-seven tons?

A. Allende
B. Ahnighito
C. Hoba West
D. El Chaco

5 Which type of meteorites are very rare?

A. Stony
B. Metallic
C. Stony-iron
D. None of the above

6 Which common element in meteors produces a bright yellow or orange glow?

A. Magnesium
B. Sodium
C. Carbon
D. None of the above

7 Where do meteoroids originate?

A. Comets
B. Asteroids
C. Collisions between celestial bodies
D. All of the above

8 What was the Hoba West Meteorite primarily composed of?

A. Silicon
B. Lead
C. Iron
D. Magnesium

Count how many times you used each letter as a correct answer (ABCD) to determine the 4-digit code. Record your answer in the boxes below.

# of A's [ ]   # of B's [ ]   # of C's [ ]   # of D's [ ]

ECLIPSES

SOLAR ECLIPSE

A solar eclipse occurs when the Moon passes directly between the Earth and the Sun, casting a shadow on Earth and partially or completely blocking the Sun's light for viewers in certain locations. This alignment, known as syzygy, occurs only during a new moon and requires precise positioning of the three celestial bodies.

Solar eclipses have intrigued and puzzled humans throughout history, often interpreted as omens of doom or significant events. The Chinese were among the first to document eclipses. In their mythology, eclipses symbolized celestial dragons devouring the Sun. During 3rd century BCE, Aristarchus of Samos used eclipses to estimate the distances between Earth, Moon, and Sun.

English scientist Edmund Halley is credited with successfully predicting and observing the solar eclipse on May 3, 1715. Halley's prediction was groundbreaking because it was based on his understanding of the mechanics of the solar system and the motion of celestial bodies. By using mathematical calculations, he was able to predict the exact timing and path of the eclipse, a remarkable achievement for the time.

The 1919 Eclipse was observed by Sir Arthur Eddington, this eclipse offered proof of Einstein's theory of general relativity. Observations during the eclipse showed that starlight bent around the Sun's gravitational field, validating the concept of space-time curvature.

Types of Solar Eclipses

Solar eclipses can be classified into four main types based on the alignment of the Earth, Moon, and Sun. A total solar eclipse happens when the Moon completely covers the Sun as seen from Earth. During this event, the Sun's outer atmosphere, the corona, becomes visible. Totality, the period of complete coverage, can last up to 7.5 minutes, though most total eclipses last less than 3 minutes. A total eclipse happens roughly every eighteen months. As the umbra reaches Earth, it narrows, creating a small region where the Sun is completely obscured by the Moon. This narrow zone is known as the zone of totality, and it is the only area on Earth where a total solar eclipse can be observed.

In a partial solar eclipse, the Moon only covers part of the Sun's disk. This occurs when the alignment is imperfect, and it is the most common type of solar eclipse.

An annular eclipse occurs when the Moon is at its farthest point (apogee) in its orbit and appears smaller than the Sun. This creates a "ring of fire" effect, where the Sun forms a bright ring around the Moon. The antumbra is the area of shadow beyond the umbra where this effect is observed.

Hybrid eclipses are rare and shift between total and annular eclipses depending on the observer's location. This occurs due to Earth's curvature and the varying distances of the Moon from Earth during its orbit.

Causes of Solar Eclipses

Solar eclipses are possible because of a cosmic coincidence: the Sun is about four-hundred times larger than the Moon, but it is also about four-hundred times farther away. This results in the Moon and Sun appearing roughly the same size in the sky. When their orbits align perfectly during a new moon, the Moon's shadow falls on Earth, creating a solar eclipse.

The Moon's orbit is tilted by about five degrees relative to Earth's orbit around the Sun. Solar eclipses can only occur when the Moon's orbit intersects the Earth's orbital plane at points called nodes. The umbra is the darkest part of the Moon's shadow, where the Sun is completely obscured. Observers within the umbra experience a total solar eclipse. The penumbra is the outer part of the shadow, where the Sun is partially visible, resulting in a partial eclipse. The antumbra refers to the area of Earth's shadow where an observer can witness an annular eclipse, rather than a total eclipse.

Occurrences

Solar eclipses occur, on average, between two and five times each year. At least one eclipse will occur during each eclipse season, which is a period when the Sun, Earth, and Moon align in a way that allows an eclipse to happen. These eclipse seasons happen roughly every six months, lasting about 34 to 38 days.

While most years see just two or three solar eclipses, there are years when five solar eclipses occur. This happens when there are both a partial eclipse and a total eclipse in the same year, sometimes with an additional annular eclipse (where the Moon's shadow does not completely cover the Sun). For example, the years 1693, 1758, 1805, 1823, 1870, and 1935 each saw five solar eclipses.

The average number of solar eclipses per century is approximately 240, a figure that includes both total and partial eclipses. These eclipses are not evenly spaced but occur due to complex orbital mechanics, with the path of totality moving across different regions of the Earth each time.

TRUE OR FALSE

After reading about the **Solar Eclipse**, read each statement below and determine if it is true or false. If the statement is true, color the coin that corresponds with that question. If the statement is false, cross out that coin value. When you are finished, add the TOTAL of **ALL TRUE** coin values to reveal a 4-digit code. One digit of the code has been provided for you. If the total is 625, a 6 would go in the first box, the 2 in the second box and so on.

A. A partial solar eclipse is the most common type of solar eclipse.

B. A total eclipse happens roughly every two months.

B
25

C. The Sun is about four-hundred times larger than the Moon.

D. The average number of solar eclipses per century is approximately 36.

E. Edmund Halley is credited with successfully predicting and observing the solar eclipse on May 3, 1715.

C
50

F. A total solar eclipse happens when the Moon completely covers the Sun as seen from Earth.

D
100

G. The corona is the darkest part of the Moon's shadow.

H. Solar eclipses can be classified into eleven main types.

After shading the coins based on your answer, add the value of ALL TRUE statements to get the final total. Record your answer in the boxes below.

DOUBLE PUZZLE

After reading about the **Solar Eclipse**, determine the word that corresponds with the statements provided below. Spell the corresponding word in the boxes to the right. You may or may not use all squares provided for each answer. Any numerical answers must be spelled out. Next, use the numbers **under** indicated letters to reveal a secret word.

1 The area of shadow beyond the umbra

2

2 The years 1693, 1758, 1805, 1823, 1870, and 1935 each saw ___ (#) solar eclipses

7

3 The Sun's outer atmosphere

4

4 The Moon's orbit is tilted by about ___ degrees relative to Earth's orbit around the Sun

5 Farthest point in orbit

1

6 Last name of scientist credited with successfully predicting the solar eclipse on May 3, 1715

7 The Sun is about ___-hundred times larger than the Moon

5 6

8 Name for alignment of three celestial bodies (Earth, Moon, Sun) during a new moon

3

9 Eclipse seasons happen roughly every ___ months

10 In Chinese mythology, eclipses symbolized celestial ___ devouring the Sun

8

SECRET WORD

1 2 3 4 5 6 7 8

LUNAR ECLIPSE

Lunar eclipses occur during a full moon when the Earth, Sun, and Moon align in such a way that Earth's shadow falls on the Moon. Depending on the alignment, a lunar eclipse can be partial, total, or penumbral. Unlike solar eclipses, lunar eclipses are safe to view with the naked eye, making them accessible to a wide audience.

Historically, these celestial phenomena have been tied to mythology, folklore, and religion, often interpreted as omens of significant events. Ancient civilizations used them to study the Moon. In ancient Egyptian mythology, lunar eclipses were closely associated with the god Set, who represented chaos, storms, and darkness. Greek astronomer Aristotle used lunar eclipses to support his argument that Earth was spherical, noting the curved shadow Earth cast on the Moon.

During the Middle Ages, lunar eclipses were often seen as predictions of natural disasters. By the 17th century, the invention of the telescope allowed scientists like Johannes Kepler to make more precise predictions about lunar eclipses and understand their mechanics.

Indigenous peoples across the world have long viewed lunar eclipses as profound celestial events that symbolize transformation, renewal, or communication with the spiritual realm. These events were often marked by ceremonies and rituals that reflected a deep connection to the natural world and the cosmos. For many Indigenous cultures, the alignment of the Earth, Sun, and Moon during an eclipse was interpreted as a moment of balance or disruption in cosmic harmony, requiring respect and acknowledgment.

Phases of a Lunar Eclipse

A lunar eclipse progresses through several phases, determined by the extent of Earth's shadow covering the Moon. The two primary parts of Earth's shadow are the umbra (the darkest, central part) and the penumbra (the lighter, outer shadow).

A penumbral eclipse occurs when the Moon passes through the penumbra, causing a subtle dimming. Penumbral lunar eclipses are relatively common, occurring about one to four times per year. However, many are not visually dramatic, as only a fraction of the penumbra might cover the Moon.

A partial eclipse is when part of the Moon enters the umbra. A dark shadow begins to cover the lunar surface, creating a striking contrast. The event can last several hours, with the Moon appearing to have a distinct "bite" taken out of its surface. Over the course of a century, there are approximately ninety partial lunar eclipses.

A total lunar eclipse is when the Moon is fully engulfed in the umbra, it takes on a reddish hue, often called the "Blood Moon." This occurs because Earth's atmosphere scatters blue light, allowing only red and orange wavelengths to reach the Moon. This process is known as Rayleigh scattering. Total lunar eclipses are relatively rare events, occurring about once every two to three years, lasting for just a few hours.

A Maximum Eclipse is the midpoint of the total eclipse, when the Moon is at its deepest within Earth's shadow. A maximum eclipse refers to the point during an eclipse when the alignment between the Earth, Moon, and Sun is closest to perfect, resulting in the greatest extent of coverage or shadow.

Ending phases of eclipses occur when the sequence reverses as the Moon gradually exits the umbra and then the penumbra.

Appearance and Viewing Conditions

During a total lunar eclipse, the Moon's reddish tint can vary in intensity depending on atmospheric conditions. For instance, volcanic eruptions or wildfires that increase particulate matter in the atmosphere can intensify the red coloration. Clear skies are essential for optimal viewing, and binoculars or telescopes enhance the details of the Moon's surface during the event.

Lunar eclipses are visible from anywhere on Earth where the Moon is above the horizon during the eclipse. Unlike solar eclipses, they are safe to view with the naked eye.

The Saros Cycle

Astronomers can predict lunar eclipses centuries in advance using calculations based on the Saros cycle, a period of approximately 18 years, 11 days, during which similar eclipses occur. The Saros cycle occurs because of the near alignment of three lunar cycles. The synodic month is the time between successive new moons. The draconic month is the time for the Moon to return to the same node of its orbit. The nodes are the two points where the Moon’s orbit intersects the ecliptic plane (the Sun’s apparent path across the sky). The anomalistic month refers to the time it takes for the Moon to return to its perigee. Perigee is the closest point to Earth in its orbit. Each lunar eclipse in a Saros cycle will have a very similar geometry, meaning it will be visible from approximately the same location on Earth.

PARAGRAPH CODE

After reading about the **Lunar Eclipse**, head back to the reading and number ALL the paragraphs in the reading passage. Then, read each statement below and determine which paragraph **NUMBER** the statement can be found in. Paragraph numbers MAY be used more than one time or not at all. Follow the directions below to reveal the 4-digit code.

A Ending phases of eclipses occur when the sequence reverses as the Moon gradually exits the umbra and then the penumbra.

B During the Middle Ages, lunar eclipses were often seen as predictions of natural disasters.

C A Maximum Eclipse is the midpoint of the total eclipse, when the Moon is at its deepest within Earth's shadow.

D Penumbral lunar eclipses are relatively common, occurring about one to four times per year.

E Depending on the alignment, a lunar eclipse can be partial, total, or penumbral.

F Lunar eclipses are visible from anywhere on Earth where the Moon is above the horizon during the eclipse.

G Historically, these celestial phenomena have been tied to mythology, folklore, and religion, often interpreted as omens of significant events.

H A partial eclipse is when part of the Moon enters the umbra.

ELIMINATE ALL EVEN-NUMBERED paragraphs that you used as an answer. Record the remaining numbers (in the SAME order in which you recorded them above) in the boxes below.

MYSTERY WORD

After reading about the **Lunar Eclipse**, determine if each statement below is true or false. Color or shade the boxes of the **TRUE** statements. Next, unscramble the mystery word using the large letters of the **TRUE** statements.

Penumbral lunar eclipses occur one to four times per year. **T**	The Saros cycle is a period of approximately 18 years, 11 days. **N**	The perigee month is the time between successive new moons. **B**	A partial eclipse is when part of the Moon enters the umbra. **C**
Lunar eclipses were associated with the god Apollo in Egyptian mythology. **L**	Over the course of a century, there are about nine-hundred partial lunar eclipses. **J**	A total lunar eclipse is often called the "Blood Moon." **R**	The Saros cycle occurs because of the near alignment of sixteen lunar cycles. **H**
Aristotle used lunar eclipses to support his argument that Earth was spherical. **E**	The penumbra is the darkest, central part of the Sun. **A**	Lunar eclipses occur during a full moon when the Earth, Sun, and Moon align. **D**	Lunar eclipses are visible from anywhere on Earth where the Moon is above the horizon. **O**
Ending phases of eclipses occur when the sequence reverses. **P**	Perigee is the closest point to Earth in its orbit. **I**	The synodic month refers to the time it takes for the Moon to return to its perigee. **F**	Total lunar eclipses occur about once every two to three years. **I**

Unscramble the word using the large bold letters of only the **TRUE** statements.

SPACE EXPLORATION

NASA SPACE PROGRAM

The National Aeronautics and Space Administration (NASA) was established in 1958 by President Dwight D. Eisenhower. Today, NASA is part of the federal government that explores space and aeronautics; the science of traveling through air. Over the years, NASA has had more than 1,000 unmanned missions and over one-hundred missions with astronauts onboard. These missions have accomplished a great deal from orbits, sending people to the Moon, and launching satellites for research.

Establishment of NASA

Before NASA was established, President Woodrow Wilson started the National Advisory Committee for Aeronautics (NACA) in 1915. NACA was created during World War I in an attempt for the U.S. government to catch up with Europe's advancements in aviation. NASA began one year after the Soviet Union (now Russia) launched the first man-made satellite, known as Sputnik, into space.

At the time, the Soviet Union and the United States were competing in the Cold War to become the global superpower. In an effort to organize their own space program, and to become more powerful than the Soviet Union, the U.S. passed the National Aeronautics and Space Act.

Project Mercury

The first successful manned mission of NASA was part of Project Mercury. Project Mercury had a few goals; the main goal was to send the first manned spacecraft to orbit Earth. They wanted to learn how to fly into space and come back safely as well as understand how humans would function in space. Many test flights with animals were completed before astronauts were sent to space.

The Mercury program began in 1958 and ended in 1963, consisting of six total flights with astronauts. Two of the six only went into space and came back to Earth, without orbiting. Four were able to orbit Earth before coming back safely.

In 1961, Russian astronaut Yuri Gagarin became the first man to orbit the Earth aboard the Vostok I. He orbited (flew around) the Earth in 89 minutes. Three weeks later Alan Shepherd became the first American in space on the Freedom 7, though he did not orbit Earth. He reached a height of 116 miles. In 1962, John Glenn became the first American astronaut to orbit Earth aboard the Friendship 7. Glenn orbited the Earth three times.

Project Gemini

In 1961, President John F. Kennedy announced that America should put a man on the Moon within the next ten years, leading to Project Gemini. Project Gemini became NASA's second manned space flight program. The goal of Gemini was to learn techniques and gather information to prepare for a successful Moon landing.

In 1966, Armstrong was asked to be the commanding pilot of the Gemini 8. The mission was to attempt the first-ever docking of two spacecrafts: the Gemini 8 and the Agena. The docking mission was successful; however, the craft began to malfunction shortly after, spinning out of control. Neil manually disabled the thrusters and activated the re-entry control system. He was able to stabilize the spacecraft, but NASA cut the mission short so that the astronauts would remain safe.

Apollo Program

In 1969, Apollo 11 was the first mission in the world to successfully put a man on the Moon. Neil Armstrong commanded the flight and first stepped foot on the Moon. He was followed by astronaut Buzz Aldrin. The Apollo program continued until 1972 and put twelve more astronauts on the Moon through five other launches.

The International Space Station (ISS) is a large spacecraft that orbits Earth. It is both a lab and a home to astronauts. NASA and other countries such as Japan and Australia worked together to build it. The ISS is made up of multiple pieces and the first piece was launched into space by a rocket in 1998. In 2000, more pieces were added making it ready for astronauts. The ISS is used to learn about what happens to humans when living in space as well as conduct experiments in biology, physics, astronomy, and meteorology. Research done in the space station cannot be done on Earth.

NASA's Organization

With headquarters in Washington DC, NASA is divided into five directories: Aerospace Technology, Space Science, Human Exploration and Development of Space, Earth Science, and Biological and Physical Research. Aerospace Technology focuses on the development of air and space transportation. The Space Science directory studies the universe, its origin, evolution, and structure. The Human Exploration and Development of Space oversees the ISS. The Earth Science directory studies the Earth. Finally, Biological and Physical Research studies the effects of space on humans. President Donald Trump established the Space Force in 2019.

MYSTERY MATCH

After reading about **NASA**, draw a line from the left-hand column to make a match in the right-hand column. Your line should go through **ONE** letter. When you complete all the matches, use the letters with a line THROUGH them to unscramble a mystery word. You MUST start and end your line at the **arrow points**.

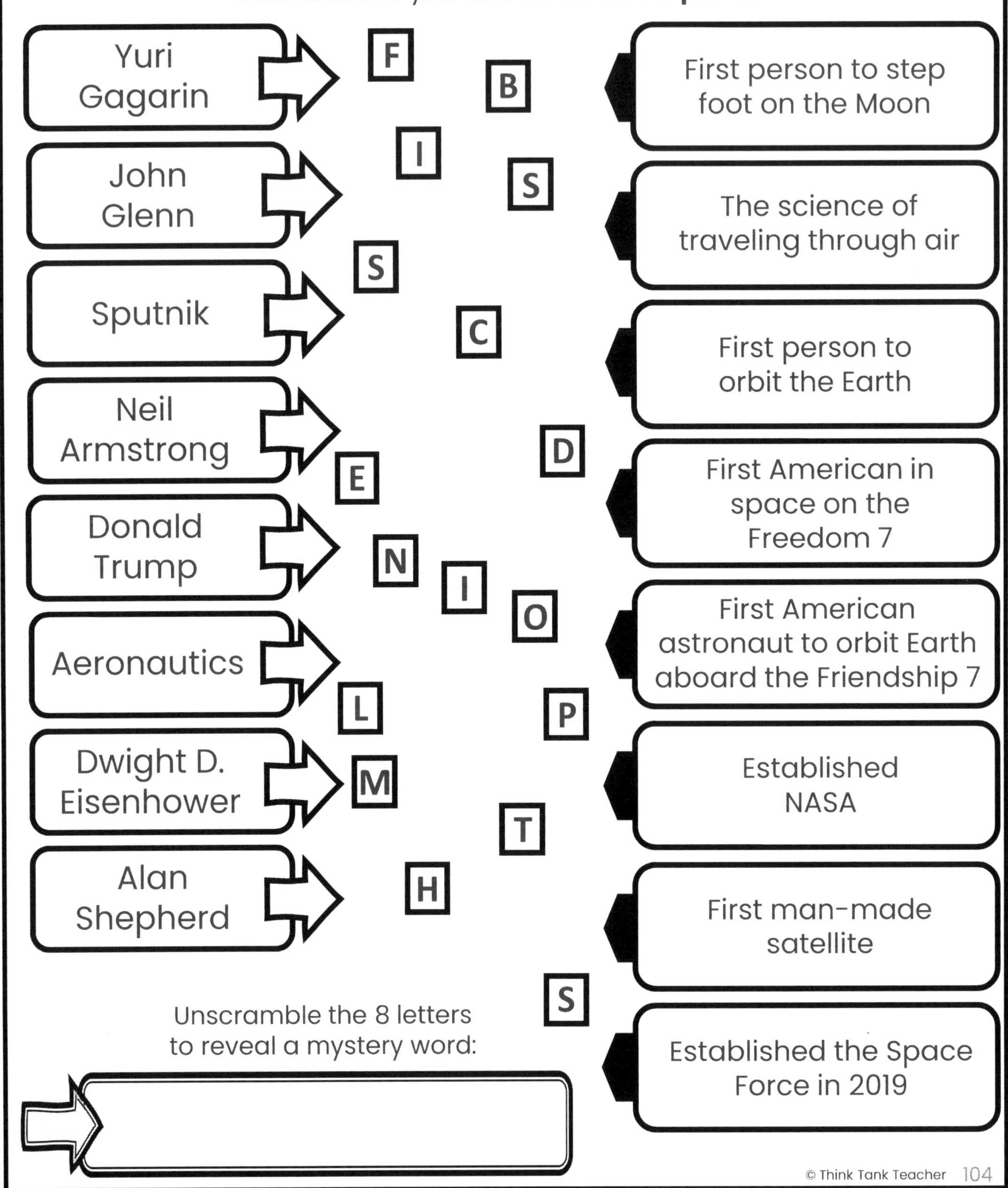

Yuri Gagarin
John Glenn
Sputnik
Neil Armstrong
Donald Trump
Aeronautics
Dwight D. Eisenhower
Alan Shepherd
F
B
I
S
S
C
D
E
N
I
O
L
P
M
T
H
S
First person to step foot on the Moon
The science of traveling through air
First person to orbit the Earth
First American in space on the Freedom 7
First American astronaut to orbit Earth aboard the Friendship 7
Established NASA
First man-made satellite
Established the Space Force in 2019

Unscramble the 8 letters to reveal a mystery word:

MULTIPLE CHOICE

After reading about **NASA**, answer each multiple-choice question below. Then, count the number of times you used each letter as an answer (ABCD) to reveal a 4-digit code. Letters may be used more than once or not at all. If a letter option is not used, put a zero in that box.

1 How many total flights with astronauts did the Mercury program consist of?

A. Six
B. Ten
C. Twelve
D. Fifteen

2 In 1969, what was the first mission in the world to successfully put a man on the Moon?

A. Friendship 7
B. Agena 3
C. Vostok I
D. Apollo 11

3 Which president started the National Advisory Committee for Aeronautics (NACA) in 1915?

A. John F. Kennedy
B. Donald Trump
C. Dwight D. Eisenhower
D. Woodrow Wilson

4 How many times did Glenn orbit the Earth aboard the Friendship 7?

A. Three
B. Five
C. Seven
D. Nine

5 When was NASA established by President Dwight D. Eisenhower?

A. 1958
B. 1963
C. 1968
D. 1971

6 The first successful manned mission of NASA was part of which project?

A. Project Agena
B. Project Apollo
C. Project Gemini
D. Project Mercury

7 In 1961, which president announced that America should put a man on the Moon?

A. John F. Kennedy
B. Donald Trump
C. Dwight D. Eisenhower
D. Woodrow Wilson

8 Russian astronaut Yuri Gagarin became the first man to orbit the Earth aboard what spacecraft?

A. Apollo 11
B. Freedom 7
C. Vostok I
D. None of the above

Count how many times you used each letter as a correct answer (ABCD) to determine the 4-digit code. Record your answer in the boxes below.

NEBULAR HYPOTHESIS

TRUE OR FALSE

After reading about the **Nebular Hypothesis**, read each statement below and determine if it is true or false. If the statement is true, color the coin that corresponds with that question. If the statement is false, cross out that coin value. When you are finished, add the TOTAL of **ALL TRUE** coin values to reveal a 4-digit code. One digit of the code has been provided for you. If the total is 625, a 6 would go in the first box, the 2 in the second box and so on.

A 75

E 100

A. Kant expanded on Swedenborg's ideas and published Universal Natural History and Theory of the Heavens in 1755.

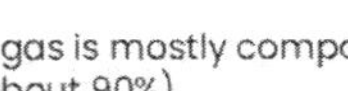

B. Interstellar gas is mostly composed of nitrogen (about 90%).

X

C. Protoplanetary disks are regions of gas and dust around young stars that resemble early solar nebula.

F 75

D. Planetary nebulae are formed from a stream of charged particles from the Sun, known as solar wind.

C 50

X

E. Nebulae form when interstellar gas and dust clump together under gravity.

F. The Nebular Hypothesis is the leading theory on the formation of our solar system.

X

H 25

G. Nebulae are often gray or yellow in color.

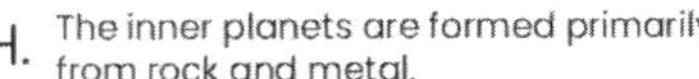

H. The inner planets are formed primarily from rock and metal.

After shading the coins based on your answer, add the value of ALL TRUE statements to get the final total. Record your answer in the boxes below.

3	2	5	2

DOUBLE PUZZLE

After reading about the **Nebular Hypothesis**, determine the word that corresponds with the statements provided below. Spell the corresponding word in the boxes to the right. You may or may not use all squares provided for each answer. Any numerical answers must be spelled out. Next, use the numbers **under** indicated letters to reveal a secret word.

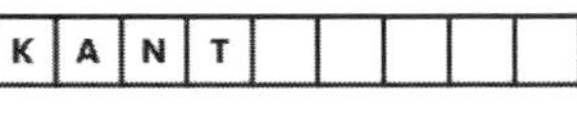

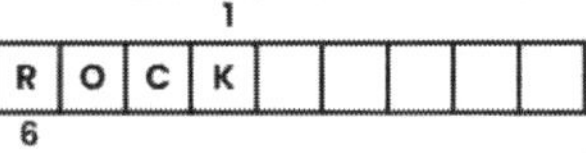

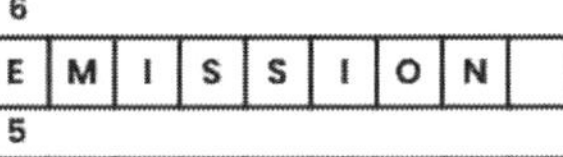

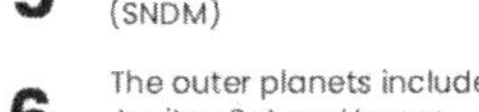

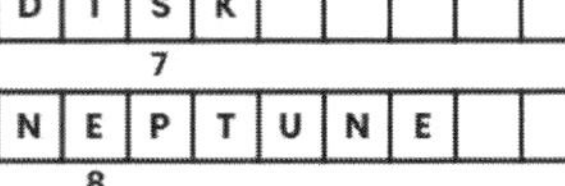

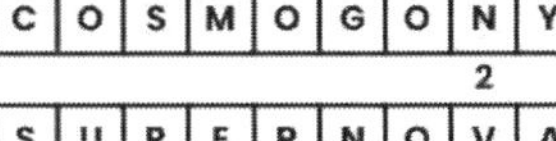

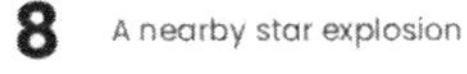

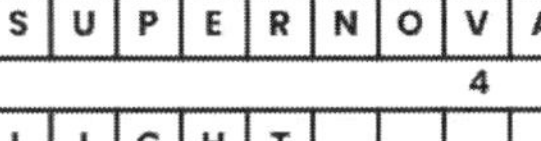

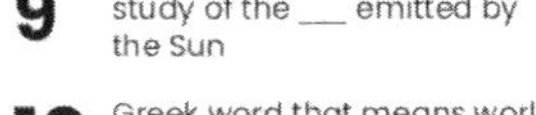

#	Clue	Answer
1	Last name of author of *Universal Natural History and Theory of the Heavens*	K A N T
2	Massive cloud of gas and dust	N E B U L A (1 under U)
3	The inner planets are formed primarily from ___ and metal	R O C K (6 under R)
4	___ nebulae glow from ionized gases	E M I S S I O N (5 under E)
5	Solar Nebula ___ Model (SNDM)	D I S K (7 under S)
6	The outer planets include Jupiter, Saturn, Uranus, and ___	N E P T U N E (8 under E)
7	The study of the origin and development of the universe	C O S M O G O N Y (2 under N)
8	A nearby star explosion	S U P E R N O V A (4 under V)
9	Solar spectroscopy is the study of the ___ emitted by the Sun	L I G H T (3 under I)
10	Greek word that means world or universe	K O S M O S

SECRET WORD

U	N	I	V	E	R	S	E
1	2	3	4	5	6	7	8

NICOLAUS COPERNICUS

PARAGRAPH CODE

After reading about **Nicolaus Copernicus**, head back to the reading and number ALL the paragraphs in the reading passage. Then, read each statement below and determine which paragraph **NUMBER** the statement can be found in. Paragraph numbers MAY be used more than one time or not at all. Follow the directions below to reveal the 4-digit code.

	Statement	Paragraph
A	The geocentric model was accepted by most people in Europe, including the Catholic Church.	5
B	It was completed around 1532 but was not published until 1543, shortly before his death.	9
C	Between 1501 and 1503, Copernicus studied at the University of Padua.	8
D	It was not until Isaac Newton came up with the Law of Universal Gravitation that the heliocentric view became widely accepted.	12
E	The evidence Copernicus provided was mathematical and theoretical, but not physical.	7
F	Copernicus was born on February 19, 1473, in Toruń, Poland.	2
G	The Catholic Church eventually placed *De Revolutionibus* on its Index of Forbidden Books in 1616.	10
H	In 1491, Copernicus enrolled at the University of Kraków where he studied mathematics, astronomy, and philosophy.	3

ELIMINATE ALL EVEN-NUMBERED paragraphs that you used as an answer. Record the remaining numbers (in the SAME order in which you recorded them above) in the boxes below.

5	9	7	3

MYSTERY WORD

After reading about **Nicolaus Copernicus**, determine if each statement below is true or false. Color or shade the boxes of the **TRUE** statements. Next, unscramble the mystery word using the large letters of the **TRUE** statements.

The insights of Copernicus sparked the Scientific Revolution. **R**	Kepler proved that planetary orbits were elliptical rather than circular. **L**	Newton's work supported the idea that the Sun was at the center of the solar system. **A**	Nicolaus Copernicus was a French astronomer. **C**
The word "helios" is an ancient Greek word meaning "moon." **H**	In 1491, Copernicus enrolled at the University of Kraków. **Y**	Copernicus hesitated to publish his work. **E**	*De Revolutionibus Orbium Coelestium* consisted of twelve volumes. **S**
Copernicus died on May 24, 1543, at the age of seventy. **A**	Copernicus could easily provide physical evidence of his theory. **O**	Copernicus argued that Earth rotated on its axis daily and orbited the Sun annually. **P**	Copernicus was born on February 19, 1476, in Madrid, Spain. **B**
Nicolaus was the oldest of four children. **D**	The geocentric model was considered a fact for many centuries. **T**	The ban of *De Revolutionibus* was lifted just three years later. **I**	The heliocentric theory initially faced resistance. **N**

Unscramble the word using the large bold letters of only the **TRUE** statements.

PLANETARY

JOHANNES KEPLER

MYSTERY MATCH

After reading about **Johannes Kepler**, draw a line from the left-hand column to make a match in the right-hand column. Your line should go through **ONE** letter. When you complete all the matches, use the letters with a line THROUGH them to unscramble a mystery word. You MUST start and end your line at the **arrow points**.

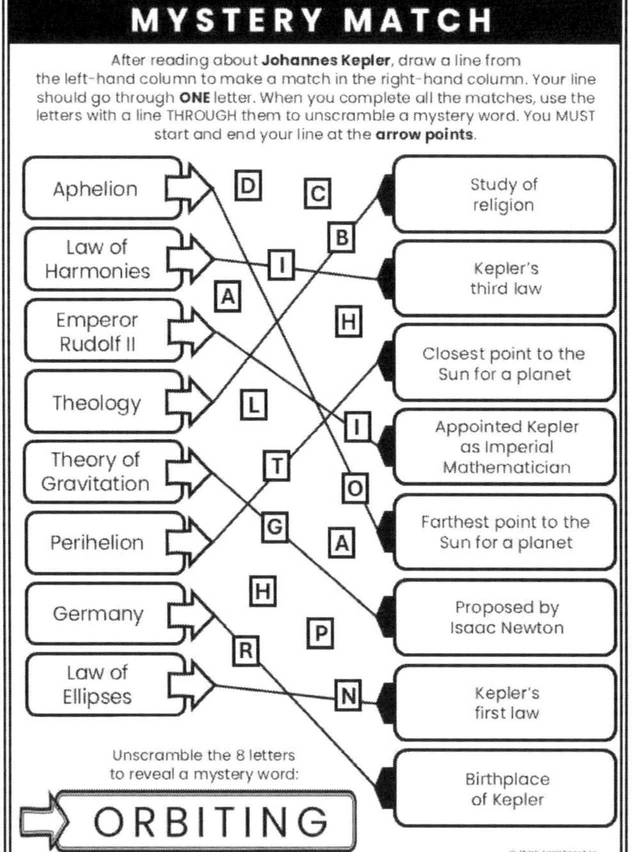

Unscramble the 8 letters to reveal a mystery word:

ORBITING

© Think Tank Teacher

MULTIPLE CHOICE

After reading about **Johannes Kepler**, answer each multiple-choice question below. Then, count the number of times you used each letter as an answer (ABCD) to reveal a 4-digit code. Letters may be used more than once or not at all. If a letter option is not used, put a zero in that box.

1 How many laws of planetary motion did Kepler formulate between 1609 and 1619?
A. One
B. Two
C. Three
D. Four

2 In which of Kepler's publications did he analyze how lenses bend light?
A. Dioptrice
B. Harmonices Mundi
C. Astronomia Nova
D. None of the above

3 Who proposed the heliocentric model?
A. Isaac Newton
B. Ptolemy
C. Tycho Brahe
D. Nicolaus Copernicus

4 What is a planet's average distance from the Sun known as?
A. Axial tilt
B. Semi-major axis
C. Concave-minor axis
D. None of the above

5 Where did Kepler join the observatory of Danish astronomer Tycho Brahe?
A. Paris
B. Prague
C. London
D. Berlin

6 Where did Kepler secure a scholarship to study?
A. Harvard University
B. University of Tübingen
C. Imperial College
D. University of Cambridge

7 When was *Astronomia Nova* published?
A. 1607
B. 1609
C. 1611
D. 1613

8 Following his death, how was Kepler honored for his contributions?
A. Telescope named after him
B. Lunar crater named after him
C. Asteroid named after him
D. All of the above

Count how many times you used each letter as a correct answer (ABCD) to determine the 4-digit code. Record your answer in the boxes below.

# of A's	# of B's	# of C's	# of D's
1	4	1	2

© Think Tank Teacher

GALILEO GALILEI

TRUE OR FALSE

After reading about **Galileo Galilei**, read each statement below and determine if it is true or false. If the statement is true, color the coin that corresponds with that question. If the statement is false, cross out that coin value. When you are finished, add the TOTAL of **ALL TRUE** coin values to reveal a 4-digit code. One digit of the code has been provided for you. If the total is 625, a 6 would go in the first box, the 2 in the second box and so on.

A. Initially, Galileo sold his telescopes to Venetian merchants to help them spot ships arriving at the port.

B. Today, Galileo is known as the "Father of our Country."

C. When Galilei attended the University of Pisa, he studied politics, hoping to become a lawyer.

D. *Du Motu* detailed Galileo's observations of the Moon and the stars.

E. Galileo discovered that Venus had phases just like Earth's moon.

F. Galileo Galilei was an astronomer, scientist, mathematician, inventor, and accomplished musician.

G. In 1608, Galileo wrote *Dialogue Concerning the Two Chief World Systems.*

H. In 1609, the telescope was invented in the Netherlands.

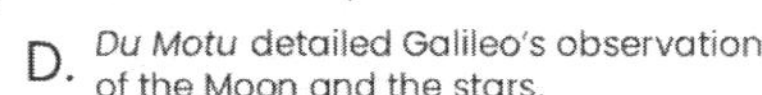

A 75 · E 100 · F 75 · H 25

After shading the coins based on your answer, add the value of ALL TRUE statements to get the final total. Record your answer in the boxes below.

2	7	5	9

© Think Tank Teacher

DOUBLE PUZZLE

After reading about **Galileo Galilei**, determine the word that corresponds with the statements provided below. Spell the corresponding word in the boxes to the right. You may or may not use all squares provided for each answer. Any numerical answers must be spelled out. Next, use the numbers **under** indicated letters to reveal a secret word.

1 Galileo was the eldest of this many children

2 Galilean moons were discovered around which planet

3 A belief that went against a religious law

4 Today, Galileo is known as the "Father of ___ Science"

5 Country where Galileo was born

6 Galileo wrote Siderius ___

7 The four objects around Jupiter were first named the Medicean ___

8 Number of Galilean moons discovered by Galileo

9 Galileo's last name

10 Galileo's telescope displayed ___ (#) times the magnification

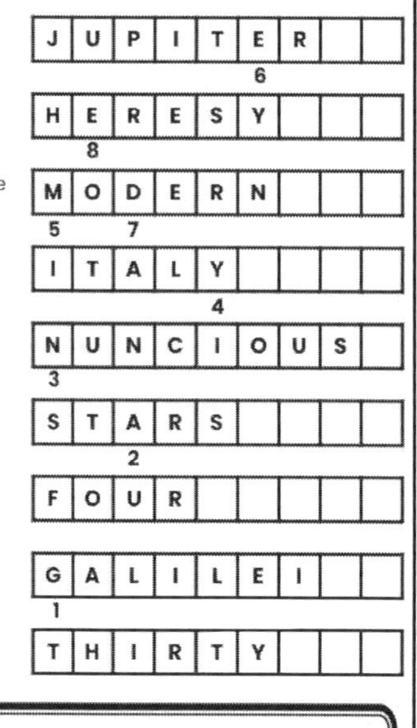

SECRET WORD

G	A	N	Y	M	E	D	E
1	2	3	4	5	6	7	8

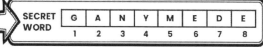

© Think Tank Teacher

THE SUN

PARAGRAPH CODE

After reading about the **Sun**, head back to the reading and number ALL the paragraphs in the reading passage. Then, read each statement below and determine which paragraph **NUMBER** the statement can be found in. Paragraph numbers MAY be used more than one time or not at all. Follow the directions below to reveal the 4-digit code.

	Statement	#
A	The outer layers include the photosphere, the chromosphere, the transition region, and the corona.	6
B	The heliosphere is the vast bubble of space dominated by the Sun's influence.	12
C	These plasma jets shoot upwards from the chromosphere into the lower corona.	7
D	Astrophysicist Arthur Eddington was the first to theorize that stars generate their energy by fusing hydrogen into helium.	8
E	The total energy the Sun emits is measured as solar luminosity.	9
F	The Sun experiences an eleven-year cycle of activity, known as the solar cycle.	10
G	The Sun is so large that it makes up 99.86% of all the mass in our solar system.	3
H	The Sun is a yellow dwarf star (G-type main-sequence star) at the center of our solar system.	1

ELIMINATE ALL EVEN-NUMBERED paragraphs that you used as an answer. Record the remaining numbers (in the SAME order in which you recorded them above) in the boxes below.

7	9	3	1

MYSTERY WORD

After reading about the **Sun**, determine if each statement below is true or false. Color or shade the boxes of the **TRUE** statements. Next, unscramble the mystery word using the large letters of the **TRUE** statements.

Surrounding the core is the radiative zone. **A**	The average time it takes for sunlight to reach Earth is about forty minutes. **S**	The Sun is predominantly composed of hydrogen and helium. **R**	Solar spicules were first observed by Albert Einstein. **L**
The solar cycle was first suggested in 1838 by astronomer Samuel Schwabe. **P**	The total energy the Sun emits is measured as solar luminosity. **E**	The Sun is so large that it makes up 99.86% of all the mass in our solar system. **T**	Photons produced in the core collide with particles in the dense radiative Zone. **R**
The Sun is a green dwarf star (Y-type main-sequence star). **B**	The Sun is composed of five inner layers and two outer layers. **C**	The Sun is a giant ball of hot gas. **E**	Every second, the Sun converts about 4.2 million metric tons of matter into energy. **M**
The Sun experiences a three-year cycle of activity, known as the solar cycle. **D**	The photosphere is the Sun's visible surface, emitting the light we see on Earth. **U**	Solar flares are intense bursts of radiation caused by magnetic energy release. **E**	Sunspots are dark, cooler areas on the Sun's surface caused by magnetic activity. **T**

Unscramble the word using the large bold letters of only the **TRUE** statements.

PHASES OF THE MOON

MYSTERY MATCH

After reading about the **Phases of the Moon**, draw a line from the left-hand column to make a match in the right-hand column. Your line should go through **ONE** letter. When you complete all the matches, use the letters with a line THROUGH them to unscramble a mystery word. You MUST start and end your line at the **arrow points**.

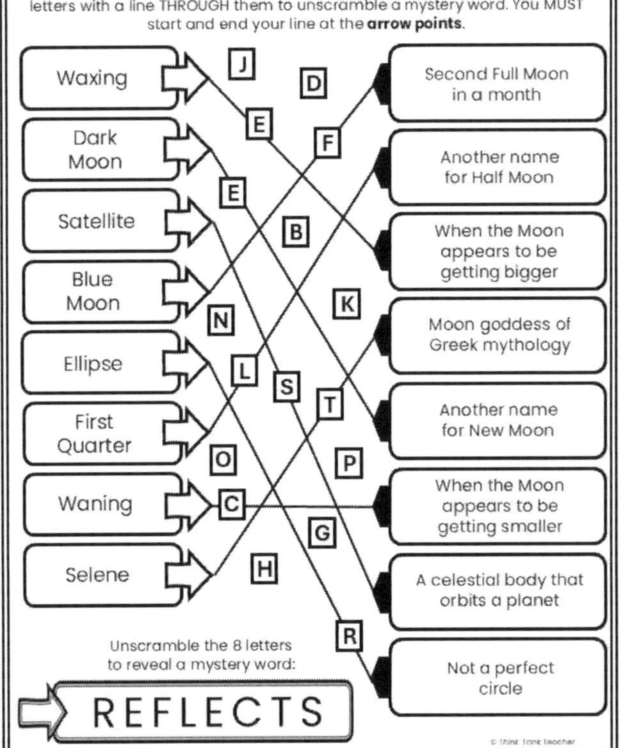

Unscramble the 8 letters to reveal a mystery word:

REFLECTS

MULTIPLE CHOICE

After reading about the **Phases of the Moon**, answer each multiple-choice question below. Then, count the number of times you used each letter as an answer (ABCD) to reveal a 4-digit code. Letters may be used more than once or not at all. If a letter option is not used, put a zero in that box.

1 What percent of the Moon is visible during the First Quarter phase?

A. 10%
B. 25%
C. 50%
D. 65%

2 In terms of moons in our solar system, where does Earth's moon rank?

A. Second largest moon
B. Third largest moon
C. Fourth largest moon
D. Fifth largest moon

3 Due to gravity, something on the Moon will be how many times lighter than on the Earth?

A. Six
B. Twelve
C. Eighteen
D. Thirty

4 What is the Moon's outermost layer composed of?

A. Iron
B. Calcium
C. Magnesium
D. All of the above

5 What is the time it takes the Moon to make a complete cycle of phases known as?

A. Lunar month
B. Lunar albedo
C. Lunar day
D. None of the above

6 Which phase occurs three or four nights after a New Moon?

A. Third Quarter
B. Waning Gibbous
C. Waxing Crescent
D. Blue Moon

7 Which ancient culture developed lunar calendars based on the phases of the Moon?

A. Chinese
B. Greeks
C. Babylonians
D. All of the above

8 How often does a Full Moon occur?

A. About every 10 days
B. About every 20 days
C. About every 30 days
D. About every 40 days

Count how many times you used each letter as a correct answer (ABCD) to determine the 4-digit code. Record your answer in the boxes below.

# of A's	# of B's	# of C's	# of D's
2	0	3	3

MERCURY

TRUE OR FALSE

After reading about **Mercury**, read each statement below and determine if it is true or false. If the statement is true, color the coin that corresponds with that question. If the statement is false, cross out that coin value. When you are finished, add the TOTAL of **ALL TRUE** coin values to reveal a 4-digit code. One digit of the code has been provided for you. If the total is 625, a 6 would go in the first box, the 2 in the second box and so on.

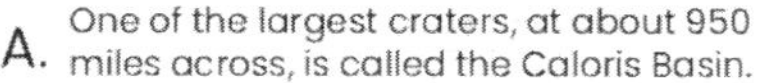

A 75 | B 25 | C 50 (crossed out) | D 100 | E 100 | F 75 | G 50 | H 25 (crossed out)

A. One of the largest craters, at about 950 miles across, is called the Caloris Basin.

B. While a year on Earth takes 365 days, it only takes 88 days for Mercury to revolve around the Sun.

C. Mercury is the largest of the eight planets in the solar system.

D. Due to the lack of atmosphere, Mercury does not experience weather, clouds, wind, or storms.

E. The time it takes for sunlight to reach Mercury is about three minutes.

F. The first four planets (Mercury, Venus, Earth, Mars) are known as the inner rocky planets, or terrestrial planets.

G. The tradition of naming craters on Mercury after famous figures began in 1976.

H. Mercury is red in color with three moons and four rings.

After shading the coins based on your answer, add the value of ALL TRUE statements to get the final total. Record your answer in the boxes below.

4	2	5	4

DOUBLE PUZZLE

After reading about **Mercury,** determine the word that corresponds with the statements provided below. Spell the corresponding word in the boxes to the right. You may or may not use all squares provided for each answer. Any numerical answers must be spelled out. Next, use the numbers **under** indicated letters to reveal a secret word.

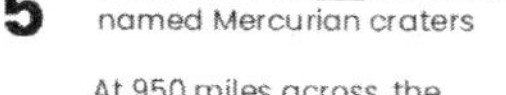
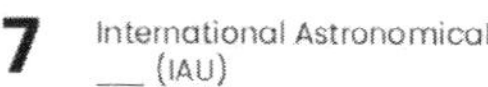

#	Statement	Answer	Number under letter
1	Terrestrial planets are also known as the ___ rocky planets	I N N E R	5 (under E)
2	Known as retrograde, Mercury appears to travel ___	B A C K W A R D	1 (under C)
3	To ancient Greeks, the planet was known as this	H E R M E S	6 (under S)
4	The first known record of Mercury was 4th century BC by ___ astronomers	A S S Y R I A N	4 (under S)
5	There are over ___ hundred named Mercurian craters	F O U R	
6	At 950 miles across, the Caloris Basin was a result of an ___	A S T E R O I D	7 (under T)
7	International Astronomical ___ (IAU)	U N I O N	
8	He was the first to see Mercury through a telescope	G A L I L E O	2 (under L)
9	In the center of Mercury, there is a large ___ core	I R O N	3 (under O)
10	The first spacecraft to visit Mercury was the ___ 10	M A R I N E R	

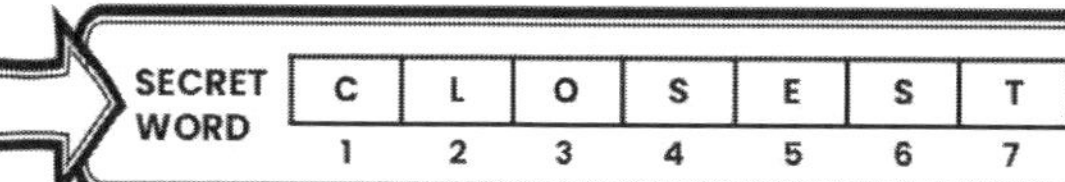

SECRET WORD

C	L	O	S	E	S	T
1	2	3	4	5	6	7

VENUS

PARAGRAPH CODE

After reading about **Venus**, head back to the reading and number ALL the paragraphs in the reading passage. Then, read each statement below and determine which paragraph **NUMBER** the statement can be found in. Paragraph numbers MAY be used more than one time or not at all. Follow the directions below to reveal the 4-digit code.

	Statement	Paragraph
A	Americans were the first on the Moon, but Russians were the first to land on Venus with the Venera 7 spacecraft.	10
B	Venus has more volcanoes than any other planet in the solar system.	7
C	Ancient civilizations such as the Babylonians, Greeks, Egyptians, and Mayans meticulously recorded the movements of Venus.	2
D	Initially, people thought Venus was two separate planets or stars: the morning star and the evening star.	3
E	Venus is the sixth largest planet in the solar system, closest planet to Earth, and the second closest planet to the Sun.	1
F	Venus has a very thick atmosphere made of mostly carbon dioxide, nitrogen, and clouds of sulfuric acid (a poisonous gas to humans).	4
G	Maxwell Montes is the highest mountain on Venus, similar in size to Mount Everest.	8
H	Venus has a rocky composition with a central core, a molten mantle, and a solid crust.	7

ELIMINATE ALL EVEN-NUMBERED paragraphs that you used as an answer. Record the remaining numbers (in the SAME order in which you recorded them above) in the boxes below.

7	3	1	7

MYSTERY WORD

After reading about **Venus**, determine if each statement below is true or false. Color or shade the boxes of the **TRUE** statements. Next, unscramble the mystery word using the large letters of the **TRUE** statements.

The mantle is a thick, silicate layer rich in magnesium and iron. **T** (shaded)	Germans were the first to land on Venus with the Magellan 7 spacecraft. **D**	Unlike Mercury, seven moons orbit Venus. **P**	Venus is the fourth closest planet to the Sun. **F**
The highland region is known as Ishtar Terra. **I** (shaded)	The morning star was called Phosphorus ("light-bringer" in Greek). **H** (shaded)	To reach Venus, the light from the Sun takes about six minutes. **E** (shaded)	Venus' mass is only about 16 percent of Earth's mass. **L**
Greeks named the planet Aphrodite after the Greek goddess of love and beauty. **R** (shaded)	Olympus Mons is the highest mountain on Venus. **N**	Even though there are clouds, it is too hot to rain on Venus. **S** (shaded)	Venus takes 243 Earth days to make one rotation. **G** (shaded)
The first successful mission to Venus took place on June 4, 1943. **A**	Venus only has two sunrises during its year. **T** (shaded)	Some of the craters on Venus look like spiders, so they are called "arachnids." **B** (shaded)	Venus is the third largest planet in the solar system. **C**

Unscramble the word using the large bold letters of only the **TRUE** statements.

BRIGHTEST

EARTH

MYSTERY MATCH

After reading about **Earth**, draw a line from the left-hand column to make a match in the right-hand column. Your line should go through **ONE** letter. When you complete all the matches, use the letters with a line THROUGH them to unscramble a mystery word. You MUST start and end your line at the **arrow points**.

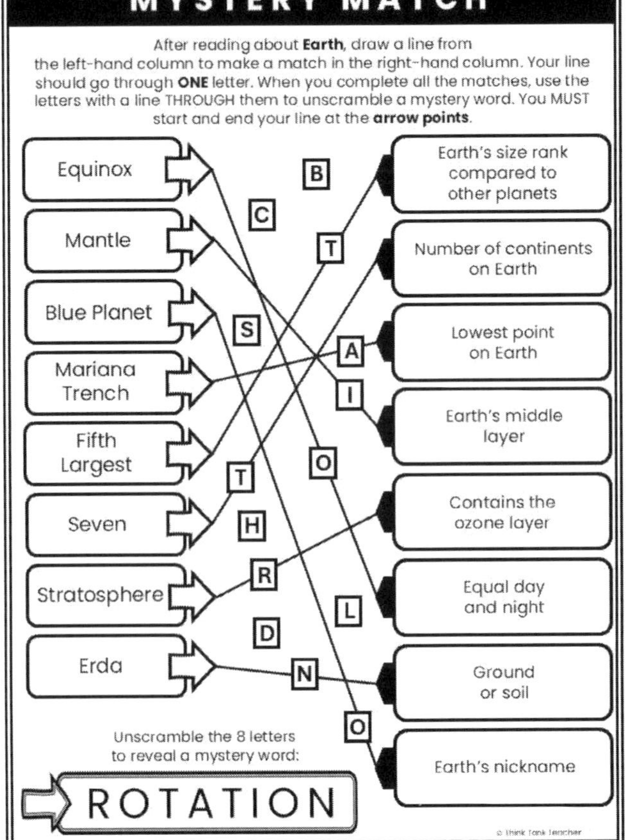

Unscramble the 8 letters to reveal a mystery word:

ROTATION

© Think Tank Teacher

MULTIPLE CHOICE

After reading about **Earth**, answer each multiple-choice question below. Then, count the number of times you used each letter as an answer (ABCD) to reveal a 4-digit code. Letters may be used more than once or not at all. If a letter option is not used, put a zero in that box.

1 What is the longest river on Earth?

A. Rio Grande River
B. Thames River
C. Nile River
D. Amazon River

2 In the solar system, how many inner terrestrial planets are there?

A. Two
B. Four
C. Six
D. Eight

3 What imaginary line from the North pole to the South pole does the Earth spin on?

A. Axis
B. Equator
C. Tropic of Cancer
D. None of the above

4 What is the outer core mainly composed of?

A. Helium and neon
B. Iodine and basalt
C. Neon and sulfur
D. Iron and nickel

5 How many main layers does Earth's atmosphere consists of?

A. Three
B. Five
C. Seven
D. Nine

6 Which ocean is also known as the Southern Ocean?

A. Antarctic
B. Pacific
C. Atlantic
D. Arctic

7 What percent of Earth's mass (weight) does the mantle make up?

A. 80%
B. 60%
C. 40%
D. 20%

8 How many moons does the Earth have in its orbit?

A. One
B. Two
C. Three
D. Four

Count how many times you used each letter as a correct answer (ABCD) to determine the 4-digit code. Record your answer in the boxes below.

# of A's	# of B's	# of C's	# of D's
4	2	1	1

© Think Tank Teacher

MARS

TRUE OR FALSE

After reading about **Mars**, read each statement below and determine if it is true or false. If the statement is true, color the coin that corresponds with that question. If the statement is false, cross out that coin value. When you are finished, add the TOTAL of **ALL TRUE** coin values to reveal a 4-digit code. One digit of the code has been provided for you. If the total is 625, a 6 would go in the first box, the 2 in the second box and so on.

A. Mars is the second largest planet in the solar system.

B. Deimos is one of the smallest moons in the solar system.

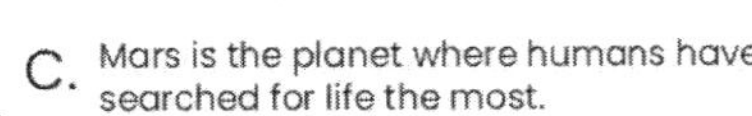

C. Mars is the planet where humans have searched for life the most.

D. Olympus Mons is nine times taller than Earth's Mt. Everest.

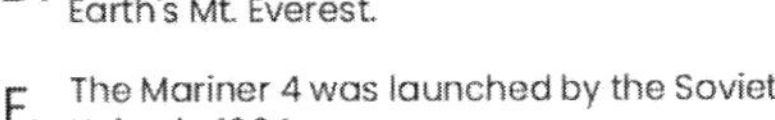

E. The Mariner 4 was launched by the Soviet Union in 1964.

F. Mars' egg-shaped orbit accounts for the difference in season lengths.

G. Mars looks red during a storm when the dust releases into the atmosphere.

H. Isaac Newton first observed Mars through a telescope in 1604.

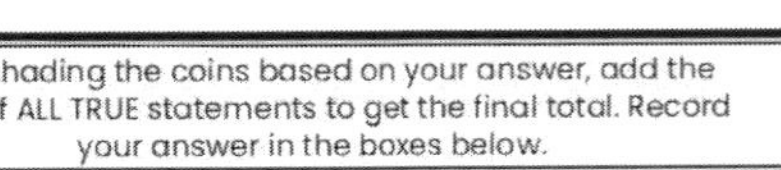

After shading the coins based on your answer, add the value of ALL TRUE statements to get the final total. Record your answer in the boxes below.

2	0	0	6

© Think Tank Teacher

DOUBLE PUZZLE

After reading about **Mars**, determine the word that corresponds with the statements provided below. Spell the corresponding word in the boxes to the right. You may or may not use all squares provided for each answer. Any numerical answers must be spelled out. Next, use the numbers **under** indicated letters to reveal a secret word.

1 The longest season on Mars

2 The moon of Mars that is closest to the planet

3 ___ Marineris is the biggest crater in the solar system

4 Mars has an ___ tilt of about 25 degrees

5 Phobos goes around Mars ___ times in one Martian day

6 Babylonians referred to the planet as this name for the god of war

7 Reddish Martian soil

8 NASA'S first successful touch down on the surface of Mars was with the ___ I lander

9 The highest mountain in the solar system is ___ Mons

10 Number of minutes it takes sunlight to reach Mars

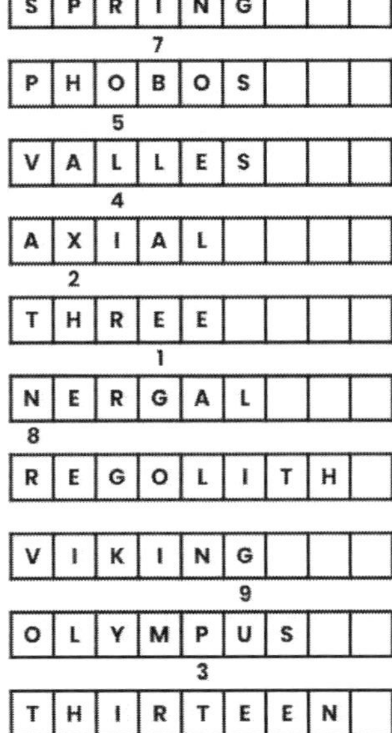

SECRET WORD

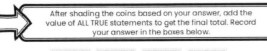

E	X	P	L	O	R	I	N	G
1	2	3	4	5	6	7	8	9

© Think Tank Teacher

JUPITER

PARAGRAPH CODE

After reading about **Jupiter**, head back to the reading and number ALL the paragraphs in the reading passage. Then, read each statement below and determine which paragraph **NUMBER** the statement can be found in. Paragraph numbers MAY be used more than one time or not at all. Follow the directions below to reveal the 4-digit code.

	Statement	#
A	Jupiter has an interesting "Great Red Spot" that Giovanni Cassini, an Italian astronomer, spotted in 1665.	5
B	Jupiter was named after the king of the Roman gods in ancient mythology.	2
C	Jupiter has four faint rings called the Halo ring, Main ring, Amalthea Gossamer ring, and the Thebe Gossamer ring.	6
D	This discovery provided evidence against the geocentric model (Earth-centered universe).	8
E	Ganymede is the ninth largest object and the largest moon in the solar system (larger than Mercury).	9
F	You could fit eleven Earths side by side within Jupiter's diameter.	3
G	Jupiter is a gas planet, like Saturn, Uranus, and Neptune, made up of 90% hydrogen and 10% helium.	4
H	Winds within the Great Red Spot can reach speeds of up to four hundred miles per hour.	5

ELIMINATE ALL EVEN-NUMBERED paragraphs that you used as an answer. Record the remaining numbers (in the SAME order in which you recorded them above) in the boxes below.

5	9	3	5

MYSTERY WORD

After reading about **Jupiter**, determine if each statement below is true or false. Color or shade the boxes of the **TRUE** statements. Next, unscramble the mystery word using the large letters of the **TRUE** statements.

Jupiter has over twenty faint rings. **H**	Jupiter is a gas planet, made up of 90% oxygen and 10% hydrogen. **O**	The Galileo spacecraft reached Jupiter in 1923. **N**	Sunlight reaches Jupiter in about 45 minutes. **T**
Jupiter rotates slower than any other planet. **B**	Io is the densest moon in the solar system. **A**	Jupiter is the largest planet in the solar system. **C**	Jupiter has the brightest auroras in the solar system. **F**
Jupiter shrinks two centimeters every year because it radiates intense heat. **P**	A day on Jupiter is only ten Earth hours. **R**	Days on Jupiter are the longest of all of the eight planets in the solar system. **G**	Callisto is Jupiter's second largest moon. **S**
Jupiter's mass is over 318 times bigger than Earth. **C**	The "Great Red Spot" was discovered by Giovanni Cassini. **E**	Galileo named the moons Europa, Ganymede, Callisto, and Io. **A**	Europa is the largest moon in the solar system. **I**

Unscramble the word using the large bold letters of only the **TRUE** statements.

SATURN

MYSTERY MATCH

After reading about **Saturn**, draw a line from the left-hand column to make a match in the right-hand column. Your line should go through **ONE** letter. When you complete all the matches, use the letters with a line THROUGH them to unscramble a mystery word. You MUST start and end your line at the **arrow points**.

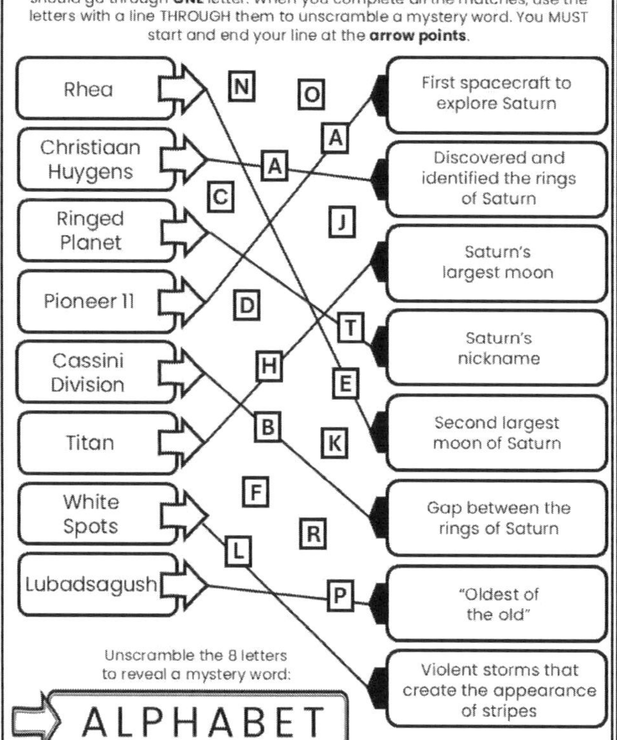

Unscramble the 8 letters to reveal a mystery word:

ALPHABET

MULTIPLE CHOICE

After reading about **Saturn**, answer each multiple-choice question below. Then, count the number of times you used each letter as an answer (ABCD) to reveal a 4-digit code. Letters may be used more than once or not at all. If a letter option is not used, put a zero in that box.

1 How many Earth years is one year on Saturn?

A. 4 years
B. 16 years
C. 29 years
D. 43 years

2 The E ring is primarily composed of ice particles ejected from which moon?

A. Dione
B. Titan
C. Mimas
D. Enceladus

3 Which moon is the largest inner moon and Saturn's fourth largest moon?

A. Rhea
B. Dione
C. Titan
D. Mimas

4 Saturn was named after the Roman god of what?

A. Sun and sky
B. Harvest and agriculture
C. Love and beauty
D. Travel and thievery

5 Which ancient civilization was the first to record Saturn in the 8th century BC?

A. Aztec
B. Anglo-Saxons
C. Assyrians
D. None of the above

6 Due to ammonia crystals in the upper atmosphere, what color does Saturn appear to be?

A. Red
B. Green
C. Yellow
D. Blue

7 When were the Voyager 1 and Voyager 2 launched?

A. 1947
B. 1957
C. 1967
D. 1977

8 In Saturn's complex ring system, how many main rings are there?

A. Two
B. Seven
C. Fifteen
D. Over fifty

Count how many times you used each letter as a correct answer (ABCD) to determine the 4-digit code. Record your answer in the boxes below.

# of A's	# of B's	# of C's	# of D's
0	3	3	2

URANUS

TRUE OR FALSE

After reading about **Uranus**, read each statement below and determine if it is true or false. If the statement is true, color the coin that corresponds with that question. If the statement is false, cross out that coin value. When you are finished, add the TOTAL of **ALL TRUE** coin values to reveal a 4-digit code. One digit of the code has been provided for you. If the total is 625, a 6 would go in the first box, the 2 in the second box and so on.

A. Miranda was discovered by Johannes Kepler in 1643.

B. The inner rings are dark and narrow and much harder to see, while the outer rings are more brightly colored.

B 25

C. Uranus is the only planet named after a Greek god.

D. Uranus is the fourth planet from the Sun and the seventh largest in the solar system.

C 50

E. Uranus spins slower than Earth and one rotation, or day, takes 158 Earth hours.

G 50

F. Umbriel is the brightest Uranian moon.

G. The five major moons include Miranda, Ariel, Umbriel, Titania, and Oberon.

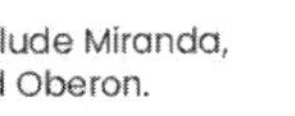

H. The discovery of Uranus doubled the size of the known solar system and challenged existing models of planetary motion.

H 25

After shading the coins based on your answer, add the value of ALL TRUE statements to get the final total. Record your answer in the boxes below.

1

5

0 7

DOUBLE PUZZLE

After reading about **Uranus**, determine the word that corresponds with the statements provided below. Spell the corresponding word in the boxes to the right. You may or may not use all squares provided for each answer. Any numerical answers must be spelled out. Next, use the numbers **under** indicated letters to reveal a secret word.

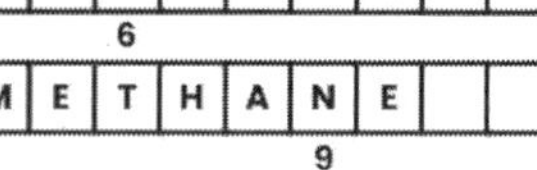

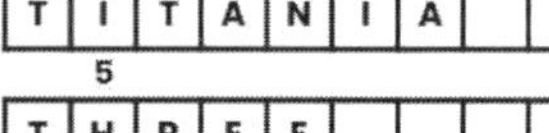

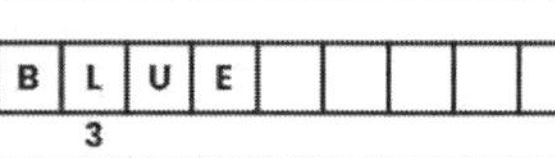

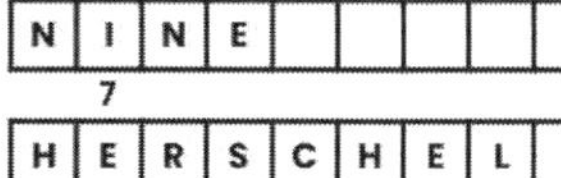

#	Clue	Answer	Number
1	Last name of astronomer that suggested the name Uranus	B O D E	8
2	The outermost of the five major moons of Uranus	O B E R O N	2
3	Brightest ring of Uranus	E P S I L O N	6
4	The third-most-abundant component of the atmosphere of Uranus	M E T H A N E	9
5	Largest moon of Uranus	T I T A N I A	5
6	Number of hours it takes sunlight to reach Uranus	T H R E E	
7	Due to the presence of methane, Uranus appears this color	B L U E	3
8	Number of moons classified as irregular moons of Uranus	N I N E	7
9	Last name of British astronomer who claimed to make ring observations in 1789	H E R S C H E L	1
10	Brightest Uranian moon	A R I E L	4

SECRET WORD

C	O	L	L	I	S	I	O	N
1	2	3	4	5	6	7	8	9

NEPTUNE

PARAGRAPH CODE

After reading about **Neptune**, head back to the reading and number ALL the paragraphs in the reading passage. Then, read each statement below and determine which paragraph **NUMBER** the statement can be found in. Paragraph numbers MAY be used more than one time or not at all. Follow the directions below to reveal the 4-digit code.

A Neptune is the smallest of the four gas giants which also include Jupiter, Saturn, and Uranus. 4

B Triton is the largest of the moons and orbits Neptune in the opposite direction.

7

C Winds can reach speeds greater than 1,200 mph, which is almost the same speed as the Hornet fighter jet in the U.S. Navy.

10

D Extreme pressure and temperatures mean that life cannot survive on Neptune.

6

E After the discovery of Neptune, Triton was discovered just seventeen days later.
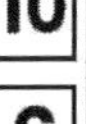
7

F The upper atmosphere of clouds contain about 80% hydrogen and 19% helium.

6

G Neptune spins, or rotates, quickly and one day takes 16 Earth hours.

5

H Neptune was named after the Roman god of the sea due to its blue appearance.

1

ELIMINATE ALL EVEN-NUMBERED paragraphs that you used as an answer. Record the remaining numbers (in the SAME order in which you recorded them above) in the boxes below.

7 7 5 1

MYSTERY WORD

After reading about **Neptune**, determine if each statement below is true or false. Color or shade the boxes of the **TRUE** statements. Next, unscramble the mystery word using the large letters of the **TRUE** statements.

Neptune is considered an ice giant along with Uranus. **E**	Neptune spins, or rotates, slowly and one day takes 681 Earth hours. **B**	The upper atmosphere of Neptune's clouds contain about 80% oxygen. **R**	Neptune is a cold, dark and windy planet. **S**
Ganymede is the second largest of Neptune's moons. **G**	Neptune's storm, called the "Great Dark Spot," lasted five years. **T**	Methane absorbs red light from the Sun. **E**	Hippocamp is the largest of Neptune's moons. **D**
Copernicus is thought to be the first person to view Neptune through a telescope in 1543. **A**	Neptune is thirty times further from the Sun than Earth is. **E**	The mantle is made up of water, ammonia, and methane ice. **O**	Neptune has fourteen official moons, with many more likely existing. **L**
Neptune was named after the Greek god of the sky because it appears blue. **U**	Neptune has the second-largest gravity of the planets. **C**	Neptune has eight rings and twelve ring arcs. **H**	There is some controversy over who discovered Neptune. **P**

Unscramble the word using the large bold letters of only the **TRUE** statements.

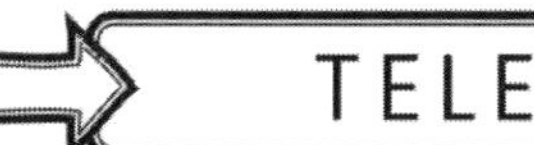
TELESCOPE

PLUTO & DWARF PLANETS

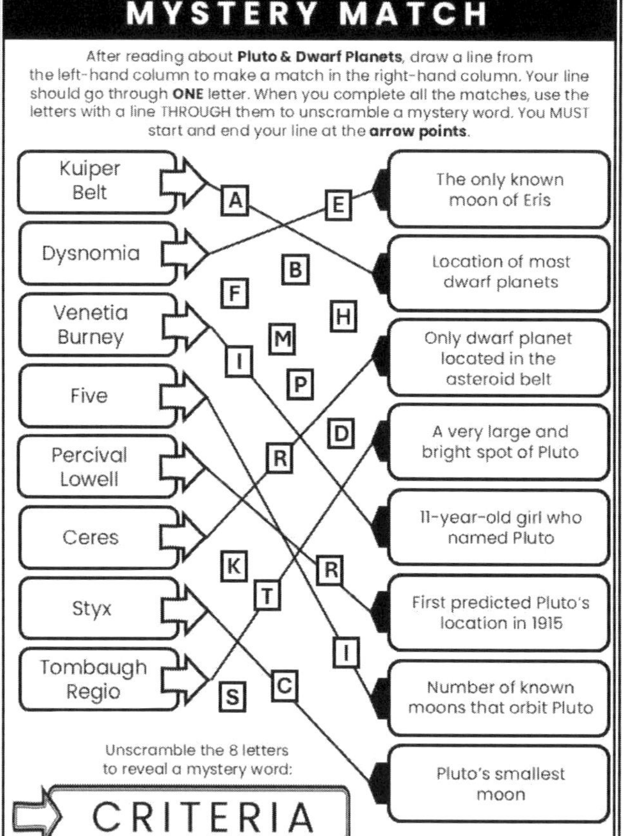

MULTIPLE CHOICE

After reading about **Pluto & Dwarf Planets**, answer each multiple-choice question below. Then, count the number of times you used each letter as an answer (ABCD) to reveal a 4-digit code. Letters may be used more than once or not at all. If a letter option is not used, put a zero in that box.

1 What covers the surface of Pluto?

A. Mountains
B. Craters
C. Valleys
D. All of the above

2 What is Pluto's largest moon?

A. Charon
B. Styx
C. Eris
D. Hydra

3 How many Earth years does it take Pluto to complete one revolution around the Sun?

A. 153 years
B. 248 years
C. 317 years
D. Over 500 years

4 When did the International Astronomical Union change Pluto's classification to a dwarf planet?

A. 2000
B. 2002
C. 2004
D. 2006

5 When was Pluto first discovered by Clyde Tombaugh?

A. 1930
B. 1940
C. 1950
D. 1960

6 Pluto was named after the Roman god of what?

A. Sea
B. Underworld
C. Sky
D. Mountains

7 Which of the following is not a dwarf planet?

A. Ceres
B. Makemake
C. Nix
D. Haumea

8 Due to hydrocarbon molecules, what color does Pluto appear?

A. Green
B. Red
C. Yellow
D. None of the above

Count how many times you used each letter as a correct answer (ABCD) to determine the 4-digit code. Record your answer in the boxes below.

# of A's	# of B's	# of C's	# of D's
2	3	1	2

ASTEROIDS

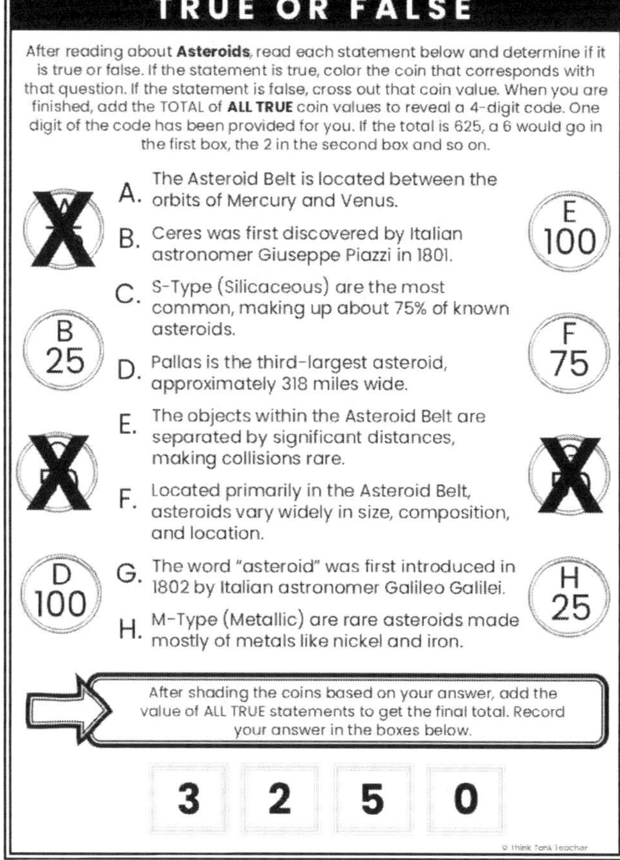

DOUBLE PUZZLE

After reading about **Asteroids**, determine the word that corresponds with the statements provided below. Spell the corresponding word in the boxes to the right. You may or may not use all squares provided for each answer. Any numerical answers must be spelled out. Next, use the numbers **under** indicated letters to reveal a secret word.

#	Clue	Answer	Number under letter
1	The brightest asteroid when viewed from Earth	VESTA	5 (under T)
2	The Asteroid Belt is located between the orbits of Mars and ___	JUPITER	
3	A small Near-Earth Asteroid (NEA) discovered in 1999	BENNU	3 (under B)
4	International Asteroid ___ Network (IAWN)	WARNING	6 (under A)
5	Asteroids are classified into ___ (#) main types	THREE	
6	___ Asteroids share an orbit with a larger planet but remain stable	TROJAN	2 (under R)
7	S-Type (Silicaceous) are also known as ___ asteroids	STONY	1 (under O)
8	Fourth-largest asteroid	HYGIEA	
9	Last name of Italian astronomer that first discovered Ceres	PIAZZI	4 (under I)
10	Asteroids are sometimes referred to as minor ___	PLANETS	7 (under L)

SECRET WORD

O	R	B	I	T	A	L
1	2	3	4	5	6	7

COMETS

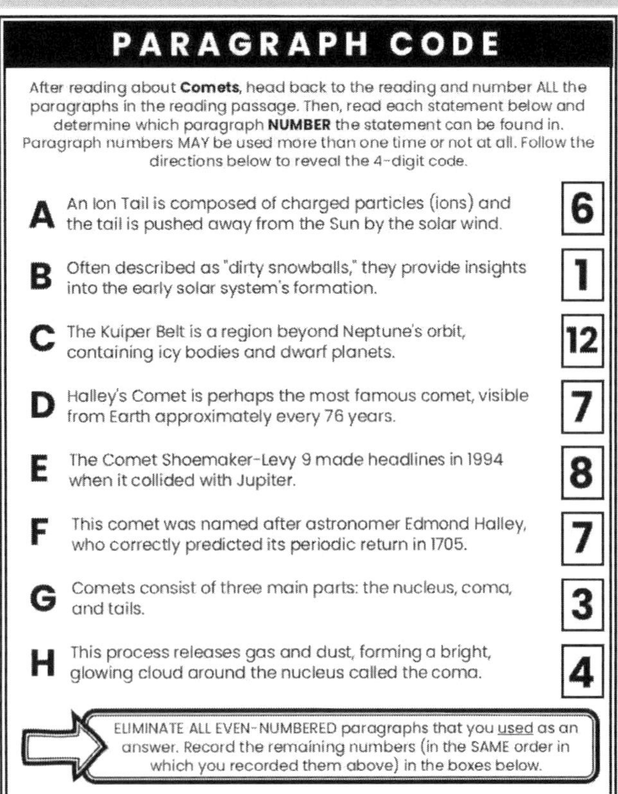

PARAGRAPH CODE

After reading about **Comets**, head back to the reading and number ALL the paragraphs in the reading passage. Then, read each statement below and determine which paragraph **NUMBER** the statement can be found in. Paragraph numbers MAY be used more than one time or not at all. Follow the directions below to reveal the 4-digit code.

	Statement	#
A	An Ion Tail is composed of charged particles (ions) and the tail is pushed away from the Sun by the solar wind.	6
B	Often described as "dirty snowballs," they provide insights into the early solar system's formation.	1
C	The Kuiper Belt is a region beyond Neptune's orbit, containing icy bodies and dwarf planets.	12
D	Halley's Comet is perhaps the most famous comet, visible from Earth approximately every 76 years.	7
E	The Comet Shoemaker-Levy 9 made headlines in 1994 when it collided with Jupiter.	8
F	This comet was named after astronomer Edmond Halley, who correctly predicted its periodic return in 1705.	7
G	Comets consist of three main parts: the nucleus, coma, and tails.	3
H	This process releases gas and dust, forming a bright, glowing cloud around the nucleus called the coma.	4

ELIMINATE ALL EVEN-NUMBERED paragraphs that you used as an answer. Record the remaining numbers (in the SAME order in which you recorded them above) in the boxes below.

1	7	7	3

MYSTERY WORD

After reading about **Comets**, determine if each statement below is true or false. Color or shade the boxes of the **TRUE** statements. Next, unscramble the mystery word using the large letters of the **TRUE** statements.

The bright, glowing cloud around the nucleus is called the oort. **T**	Comets develop four types of tails. **P**	The Hale-Bopp Comet was one of the brightest comets of the 20th century. **G**	Sublimation is when frozen gases turn directly into vapor. **L**
The nucleus is the liquid core of a comet. **B**	When a comet approaches the Sun, the heat causes rain. **F**	Short period comets always take less than fifty years to complete a full cycle. **A**	Halley's Comet is visible from Earth approximately every 76 years. **N**
Dust Tails often appear blue due to ionized gases. **E**	Comets are composed of rock, dust, ice, and frozen gases. **O**	Halley's Comet collided with Jupiter in 1994. **R**	Comets are believed to originate from two main regions in the solar system. **W**
Comets consist of six main parts. **C**	The Kuiper Belt is a region beyond Neptune's orbit. **G**	A comet's tail is formed from oxygen. **D**	Short-period comets are defined by their quick orbits around the Sun. **I**

Unscramble the word using the large bold letters of only the **TRUE** statements.

GLOWING

METEORS

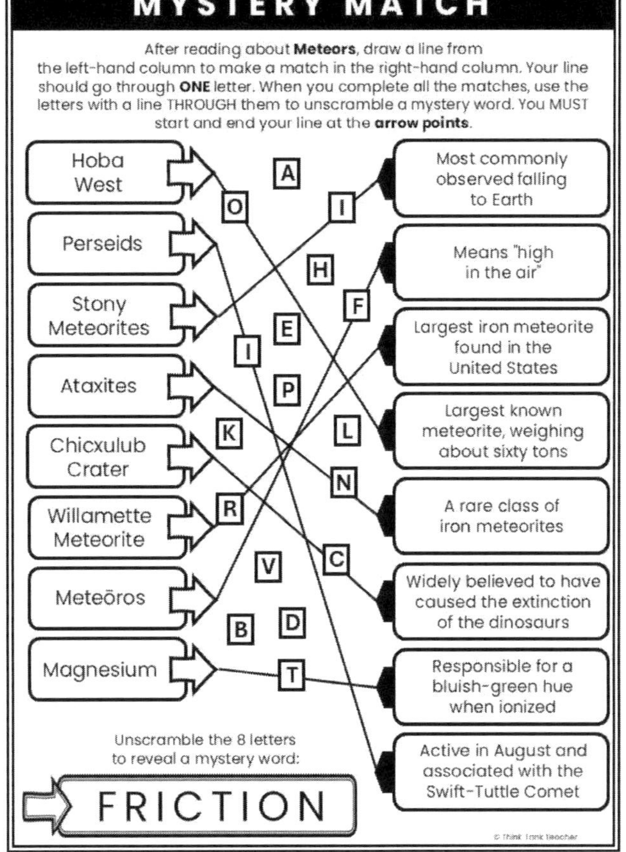

MYSTERY MATCH

After reading about **Meteors**, draw a line from the left-hand column to make a match in the right-hand column. Your line should go through **ONE** letter. When you complete all the matches, use the letters with a line THROUGH them to unscramble a mystery word. You MUST start and end your line at the **arrow points**.

Unscramble the 8 letters to reveal a mystery word:

FRICTION

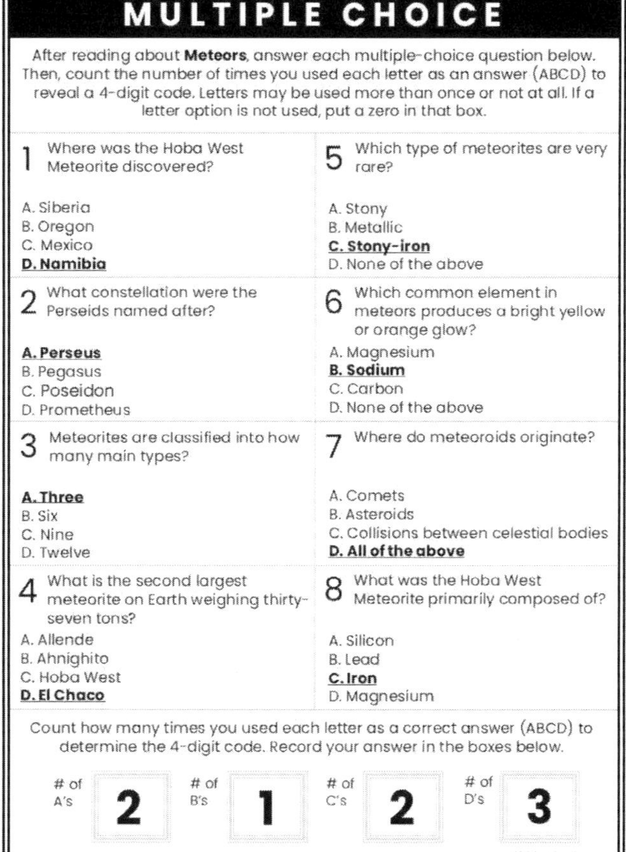

MULTIPLE CHOICE

After reading about **Meteors**, answer each multiple-choice question below. Then, count the number of times you used each letter as an answer (ABCD) to reveal a 4-digit code. Letters may be used more than once or not at all. If a letter option is not used, put a zero in that box.

1 Where was the Hoba West Meteorite discovered?

A. Siberia
B. Oregon
C. Mexico
D. Namibia

2 What constellation were the Perseids named after?

A. Perseus
B. Pegasus
C. Poseidon
D. Prometheus

3 Meteorites are classified into how many main types?

A. Three
B. Six
C. Nine
D. Twelve

4 What is the second largest meteorite on Earth weighing thirty-seven tons?

A. Allende
B. Ahnighito
C. Hoba West
D. El Chaco

5 Which type of meteorites are very rare?

A. Stony
B. Metallic
C. Stony-iron
D. None of the above

6 Which common element in meteors produces a bright yellow or orange glow?

A. Magnesium
B. Sodium
C. Carbon
D. None of the above

7 Where do meteoroids originate?

A. Comets
B. Asteroids
C. Collisions between celestial bodies
D. All of the above

8 What was the Hoba West Meteorite primarily composed of?

A. Silicon
B. Lead
C. Iron
D. Magnesium

Count how many times you used each letter as a correct answer (ABCD) to determine the 4-digit code. Record your answer in the boxes below.

# of A's	# of B's	# of C's	# of D's
2	1	2	3

SOLAR ECLIPSE

TRUE OR FALSE

After reading about the **Solar Eclipse**, read each statement below and determine if it is true or false. If the statement is true, color the coin that corresponds with that question. If the statement is false, cross out that coin value. When you are finished, add the TOTAL of **ALL TRUE** coin values to reveal a 4-digit code. One digit of the code has been provided for you. If the total is 625, a 6 would go in the first box, the 2 in the second box and so on.

A. A partial solar eclipse is the most common type of solar eclipse.

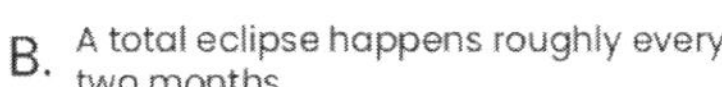

B. A total eclipse happens roughly every two months.

C. The Sun is about four-hundred times larger than the Moon.

D. The average number of solar eclipses per century is approximately 36.

E. Edmund Halley is credited with successfully predicting and observing the solar eclipse on May 3, 1715.

C 50

F. A total solar eclipse happens when the Moon completely covers the Sun as seen from Earth.

G. The corona is the darkest part of the Moon's shadow.

H. Solar eclipses can be classified into eleven main types.

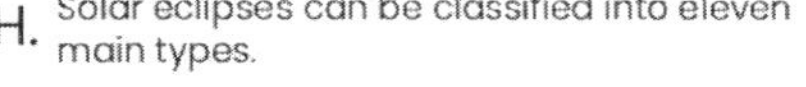

After shading the coins based on your answer, add the value of ALL TRUE statements to get the final total. Record your answer in the boxes below.

3	0	0	2

DOUBLE PUZZLE

After reading about the **Solar Eclipse**, determine the word that corresponds with the statements provided below. Spell the corresponding word in the boxes to the right. You may or may not use all squares provided for each answer. Any numerical answers must be spelled out. Next, use the numbers **under** indicated letters to reveal a secret word.

#	Clue	Answer	Numbered letters
1	The area of shadow beyond the umbra	A N T U M B R A	B = 2
2	The years 1693, 1758, 1805, 1823, 1870, and 1935 each saw ___ (#) solar eclipses	F I V E	E = 7
3	The Sun's outer atmosphere	C O R O N A	C = 4
4	The Moon's orbit is tilted by about ___ degrees relative to Earth's orbit around the Sun	F I V E	
5	Farthest point in orbit	A P O G E E	O = 1
6	Last name of scientist credited with successfully predicting the solar eclipse on May 3, 1715	H A L L E Y	
7	The Sun is about ___-hundred times larger than the Moon	F O U R	U = 5, R = 6
8	Name for alignment of three celestial bodies (Earth, Moon, Sun) during a new moon	S Y Z Y G Y	S = 3
9	Eclipse seasons happen roughly every ___ months	S I X	
10	In Chinese mythology, eclipses symbolized celestial ___ devouring the Sun	D R A G O N S	D = 8

SECRET WORD

O	B	S	C	U	R	E	D
1	2	3	4	5	6	7	8

LUNAR ECLIPSE

PARAGRAPH CODE

After reading about the **Lunar Eclipse**, head back to the reading and number ALL the paragraphs in the reading passage. Then, read each statement below and determine which paragraph **NUMBER** the statement can be found in. Paragraph numbers MAY be used more than one time or not at all. Follow the directions below to reveal the 4-digit code.

	Statement	Paragraph
A	Ending phases of eclipses occur when the sequence reverses as the Moon gradually exits the umbra and then the penumbra.	10
B	During the Middle Ages, lunar eclipses were often seen as predictions of natural disasters.	3
C	A Maximum Eclipse is the midpoint of the total eclipse, when the Moon is at its deepest within Earth's shadow.	9
D	Penumbral lunar eclipses are relatively common, occurring about one to four times per year.	6
E	Depending on the alignment, a lunar eclipse can be partial, total, or penumbral.	1
F	Lunar eclipses are visible from anywhere on Earth where the Moon is above the horizon during the eclipse.	12
G	Historically, these celestial phenomena have been tied to mythology, folklore, and religion, often interpreted as omens of significant events.	2
H	A partial eclipse is when part of the Moon enters the umbra.	7

ELIMINATE ALL EVEN-NUMBERED paragraphs that you used as an answer. Record the remaining numbers (in the SAME order in which you recorded them above) in the boxes below.

3	9	1	7

MYSTERY WORD

After reading about the **Lunar Eclipse**, determine if each statement below is true or false. Color or shade the boxes of the **TRUE** statements. Next, unscramble the mystery word using the large letters of the **TRUE** statements.

Penumbral lunar eclipses occur one to four times per year. **T**	The Saros cycle is a period of approximately 18 years, 11 days. **N**	The perigee month is the time between successive new moons. **B**	A partial eclipse is when part of the Moon enters the umbra. **C**
Lunar eclipses were associated with the god Apollo in Egyptian mythology. **L**	Over the course of a century, there are about nine-hundred partial lunar eclipses. **J**	A total lunar eclipse is often called the "Blood Moon." **R**	The Saros cycle occurs because of the near alignment of sixteen lunar cycles. **H**
Aristotle used lunar eclipses to support his argument that Earth was spherical. **E**	The penumbra is the darkest, central part of the Sun. **A**	Lunar eclipses occur during a full moon when the Earth, Sun, and Moon align. **D**	Lunar eclipses are visible from anywhere on Earth where the Moon is above the horizon. **O**
Ending phases of eclipses occur when the sequence reverses. **P**	Perigee is the closest point to Earth in its orbit. **I**	The synodic month refers to the time it takes for the Moon to return to its perigee. **F**	Total lunar eclipses occur about once every two to three years. **I**

Unscramble the word using the large bold letters of only the **TRUE** statements.

PREDICTION

NASA SPACE PROGRAM

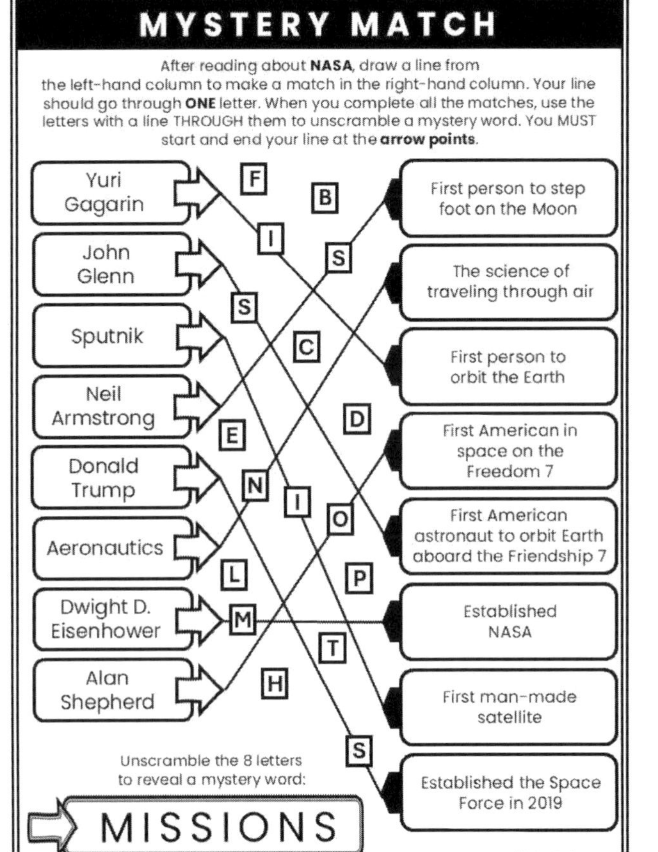

MYSTERY MATCH

After reading about **NASA**, draw a line from the left-hand column to make a match in the right-hand column. Your line should go through **ONE** letter. When you complete all the matches, use the letters with a line THROUGH them to unscramble a mystery word. You MUST start and end your line at the **arrow points**.

Unscramble the 8 letters to reveal a mystery word:

MISSIONS

© Think Tank Teacher

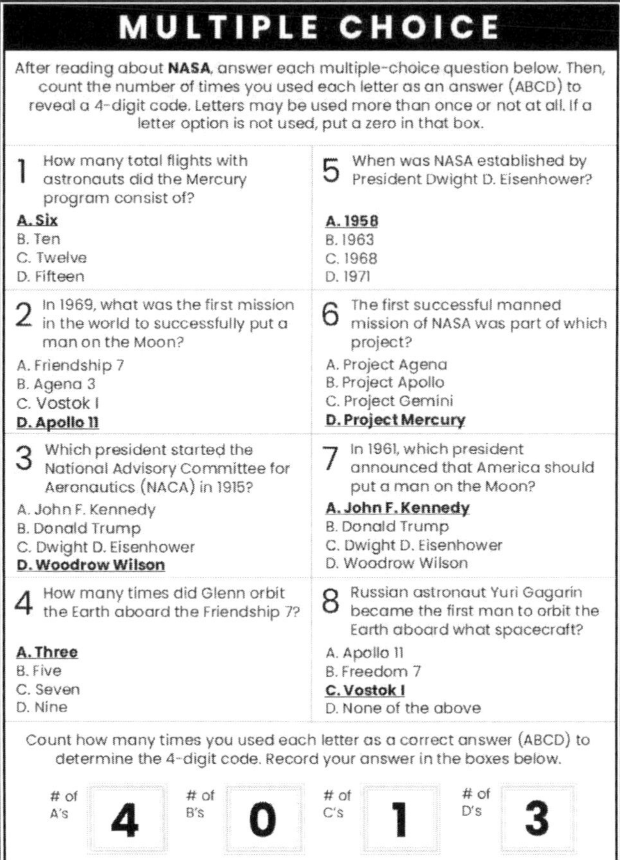

MULTIPLE CHOICE

After reading about **NASA**, answer each multiple-choice question below. Then, count the number of times you used each letter as an answer (ABCD) to reveal a 4-digit code. Letters may be used more than once or not at all. If a letter option is not used, put a zero in that box.

1 How many total flights with astronauts did the Mercury program consist of?
A. Six
B. Ten
C. Twelve
D. Fifteen

2 In 1969, what was the first mission in the world to successfully put a man on the Moon?
A. Friendship 7
B. Agena 3
C. Vostok I
D. Apollo 11

3 Which president started the National Advisory Committee for Aeronautics (NACA) in 1915?
A. John F. Kennedy
B. Donald Trump
C. Dwight D. Eisenhower
D. Woodrow Wilson

4 How many times did Glenn orbit the Earth aboard the Friendship 7?
A. Three
B. Five
C. Seven
D. Nine

5 When was NASA established by President Dwight D. Eisenhower?
A. 1958
B. 1963
C. 1968
D. 1971

6 The first successful manned mission of NASA was part of which project?
A. Project Agena
B. Project Apollo
C. Project Gemini
D. Project Mercury

7 In 1961, which president announced that America should put a man on the Moon?
A. John F. Kennedy
B. Donald Trump
C. Dwight D. Eisenhower
D. Woodrow Wilson

8 Russian astronaut Yuri Gagarin became the first man to orbit the Earth aboard what spacecraft?
A. Apollo 11
B. Freedom 7
C. Vostok I
D. None of the above

Count how many times you used each letter as a correct answer (ABCD) to determine the 4-digit code. Record your answer in the boxes below.

# of A's	# of B's	# of C's	# of D's
4	0	1	3

© Think Tank Teacher

TERMS OF USE

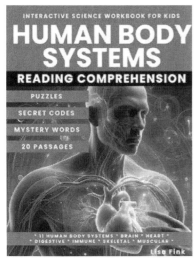

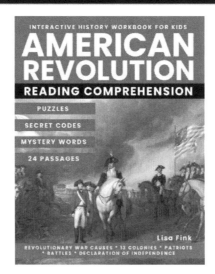

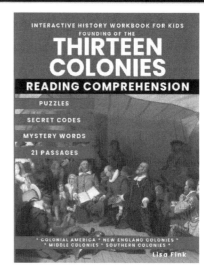

Made in the USA
Middletown, DE
14 March 2025

72688885R00066